AF411415

Applied Sciences to the Study of Technical Historical Heritage and/or Industrial Heritage

Applied Sciences to the Study of Technical Historical Heritage and/or Industrial Heritage

Special Issue Editor

José Ignacio Rojas Sola

MDPI • Basel • Beijing • Wuhan • Barcelona • Belgrade • Manchester • Tokyo • Cluj • Tianjin

Special Issue Editor
José Ignacio Rojas Sola
University of Jaen
Spain

Editorial Office
MDPI
St. Alban-Anlage 66
4052 Basel, Switzerland

This is a reprint of articles from the Special Issue published online in the open access journal *Applied Sciences* (ISSN 2076-3417) (available at: https://www.mdpi.com/journal/applsci/special_issues/historical_heritage).

For citation purposes, cite each article independently as indicated on the article page online and as indicated below:

LastName, A.A.; LastName, B.B.; LastName, C.C. Article Title. *Journal Name* **Year**, *Article Number*, Page Range.

ISBN 978-3-03936-439-8 (Hbk)
ISBN 978-3-03936-440-4 (PDF)

Contents

About the Special Issue Editor

José Ignacio Rojas Sola (Ph.D.) is a full professor at the Department of Engineering Graphics, Design and Projects, University of Jaen (Spain). He received his M.S. from the University of Seville and his Ph.D. in industrial engineering from the National University of Distance Education (UNED). He has been head of the Engineering Graphics and Industrial Archaeology research group since 1995, and was the Director for Western Europe of International Society for Geometry and Graphics from 1999 to 2008. He is a scientific reviewer for many different associations and more than 45 international journals indexed in Journal Citation Reports (JCR), and has authored more than 260 journal and conference papers, including more than 85 published in JCR journals. He has been supervisor of 11 doctoral theses, a member of the editorial boards of scientific several JCR journals (Associate Editor), and a member of the scientific committee, steering/honor committee and program committee in more than 110 international conferences. His research activities include engineering graphics applied to technical historical or industrial heritage, vernacular architecture, industrial archaeology, history of technology, computer-aided design, computer-aided engineering, and virtual and augmented reality, among others.

Preface to "Applied Sciences to the Study of Technical Historical Heritage and/or Industrial Heritage"

More initiatives from national, regional, and local administration and private institutions are being implemented to recover the legacy of technical historical and industrial heritage, since it constitutes a clear example of the technological evolution footprint that improves our understanding of the advances in societies and of the development of humanity.

The study of cultural heritage, in particular of technical historical and/or industrial heritage, is becoming increasingly important. In general, cultural heritage creates a value chain consisting of seven differentiated milestones: identification; registration and research; protection; training, conservation, and restoration; value and dissemination; heritage management; and the application of new technologies. However, this Special Issue focuses only on valorization and diffusion, as well as the use of the new technologies, concentrating on research from three main aspects: location, documentation, and dissemination.

The numerous examples that can be found are, in most cases, in a very poor state of conservation, demonstrating development ideas, or simply providing geometric documentation examples that were not preserved, necessitating its reconstruction, recovery, and study.

Thus, the different tools used for the study and recovery of technical historical and industrial heritage rely heavily on the use of engineering graphics techniques, such as geometric modelling, computer-aided design, computer-aided engineering, virtual reality and augmented reality techniques, and 3D printing, as well as knowledge of engineering and architecture in their different fields of application.

The scope of the case studies is broad, covering different engineering disciplines, such as mechanical, civil, chemical, electrical, electronic, automation and robotic, and telecommunications engineering, among others, as well as industrial architecture. In particular, research is required on historical inventions related to these fields of engineering and architecture, as well as studies of the work (historical inventions) of world-renowned engineers and architects.

Therefore, this book includes recent theoretical and practical advances in the sciences applied to the study of technical historical and/or industrial heritage related to engineering and architecture.

José Ignacio Rojas Sola
Special Issue Editor

Editorial

Applied Sciences to the Study of Technical Historical Heritage and/or Industrial Heritage

José Ignacio Rojas-Sola

Department of Engineering Graphics, Design and Projects, University of Jaén, 23071 Jaén, Spain; jirojas@ujaen.es; Tel.: +34-953-212452

Received: 1 May 2020; Accepted: 20 May 2020; Published: 25 May 2020

Abstract: Technical historical heritage and/or industrial heritage are manifestations of heritage that acquire greater relevance every day, since their study and analysis provide a global vision of their impact on the development of the societies and, also, because they favor the understanding of the technological evolution of these societies. The fields of action are very broad, both from the point of view of engineering and its different disciplines as well as from architecture. This Special Issue shows the reader some of the tools currently available to value this heritage and promote its dissemination, such as geometric modeling, computer-aided design, computer-aided engineering, and the study of industrial heritage from a global perspective.

Keywords: technical historical heritage; industrial heritage; computer-aided design; computer-aided engineering; engineering graphics; 3D printing; ancient artifacts; historical inventions

1. Introduction

The study of cultural heritage, in particular of technical historical heritage and/or industrial heritage, is becoming increasingly important. In general, cultural heritage follows a value chain consisting of seven clearly differentiated milestones: identification, registration and research; protection; training; conservation and restoration; value and dissemination; heritage management; and the application of new technologies. However, this Special Issue focuses only on the valorization and diffusion, as well as the use of, new technologies, concentrating on research in three main aspects: location, documentation and dissemination.

The scope for case studies is very broad and can cover different disciplines of engineering, such as mechanical engineering, civil engineering, chemical engineering, electrical and electronic engineering, automation and robotic engineering and telecommunications engineering, among others, as well as industrial architecture. In particular, we especially welcome research on historical inventions related to these fields of engineering and architecture, as well as studies of the work (historical inventions) of world-renowned engineers and architects.

This Special Issue invites researchers to submit original research papers and review articles related to any discipline in which the theoretical or practical issues of technical historical heritage and/or industrial heritage are considered. The topics of interest include, but are not limited to:

- Historical technical heritage
- Industrial heritage
- Industrial archaeology
- Industrial architecture
- Ancient artifacts (historical inventions)
- Computer-aided design
- Computer-aided engineering

- Geometric modeling
- Computer animation
- Virtual reality
- Augmented reality
- Engineering graphics
- Multimedia
- 3D printing

2. Statistics of the Special Issue

The statistics of the call for papers for this Special Issue, related to published or rejected items, are as follows: total submissions, 15; published, 11 (73%); rejected, 4 (17%).

The authors' geographical distribution by country for published papers is shown in Table 1, where it is possible to observe 30 authors from three different countries. Note that it is usual for an article to be signed by more than one author and for authors to collaborate with others at different affiliations.

Table 1. Geographic distribution by the country of author.

Country	Number of Authors
Spain	18
Taiwan	8
Thailand	3
Portugal	1
Total	30

3. Authors of this Special Issue

The authors of this Special Issue and their main affiliations are summarized in Table 2, where there are three authors on average per manuscript.

Table 2. Affiliations and bibliometric indicators for the authors.

Author	Main Affiliation	Reference
Yu-Hsun Chen	National Taiwan University of Science and Technology	[1]
Guan-Chen Chen	Tamkang University	[1]
Ching-Tai Wu	National Taiwan Science and Technology Museum	[1]
Chi-Lin Lee4	Tamkang University	[1]
You-Rou Chen	National Taiwan Science and Technology Museum	[1]
Jun-Fu Huang	National Taiwan Science and Technology Museum	[1]
Kuo-Hung Hsiao	National Taiwan Science and Technology Museum	[1]
Jong-I Lin	National Taiwan Science and Technology Museum	[1]
José Ignacio Rojas-Sola	University of Jaén	[2–5]
Eduardo De la Morena-De la Fuente	University de Jaén	[2,4,5]
Gloria del Río-Cidoncha	University of Sevilla	[3]
Francisco Javier González-Cabanes	University of Sevilla	[3]
Francisco García-Ahumada	Universidad Nacional de Educación a Distancia	[6]
Cristina González-Gaya	Universidad Nacional de Educación a Distancia	[6,7]

Table 2. *Cont.*

Author	Main Affiliation	Reference
Carlos J. Pardo-Abad	Universidad Nacional de Educación a Distancia	[8]
Alberto Sánchez	Universidad de Valladolid	[7]
Patricia Zulueta	Universidad de Valladolid	[7]
Zita Sampaio	Universidade de Lisboa	[7]
Beatriz Torre	Universidad de Valladolid	[7]
Supaporn Manajitprasert	Asian Institute of Technology	[9]
Nitin K. Tripathi	Asian Institute of Technology	[9]
Sanit Arunplod	Asian Institute of Technology	[9]
José Luis Saorín	University of La Laguna	[10]
Vicente López-Chao	University of La Laguna	[10]
Jorge de la Torre-Cantero	University of La Laguna	[10]
Manuel Drago Díaz-Alemán	University of La Laguna	[10]
Juan Claver	Universidad Nacional de Educación a Distancia	[11]
Amabel García-Domínguez	Universidad Nacional de Educación a Distancia	[11]
Lorenzo Sevilla	Universidad de Málaga	[11]
Miguel A. Sebastián	Universidad Nacional de Educación a Distancia	[11]

4. Brief Overview of the Contributions to This Special Issue

The analysis of the topics (Table 3) identifies or summarizes the research undertaken. This section classifies the manuscripts according to the topics proposed in the Special Issue. It was observed that five topics dominated the others: industrial heritage; geometric modeling; computer-aided design; computer-aided engineering; and historical technical heritage.

Table 3. Topic analysis.

Topic	Number of Manuscripts
Industrial Heritage	4
Geometric Modeling	2
Computer-Aided Design	2
Computer-Aided Engineering	2
Historical Technical Heritage	1
Total	11

Acknowledgments: I would like to thank all authors, the many dedicated referees, the editorial team of *Applied Sciences*, and especially Xiaoyan Chen (Managing Editor) for their valuable contributions, making this special issue a success.

Conflicts of Interest: The author declare no conflict of interest.

References

1. Chen, Y.H.; Chen, G.C.; Wu, C.T.; Lee, C.L.; Chen, Y.R.; Huang, J.F.; Hsiao, K.H.; Lin, J.I. Object investigation of industrial heritage: The forging and metallurgy shop in Taipei railway workshop. *Appl. Sci.* **2020**, *10*, 2408. [CrossRef]
2. Rojas-Sola, J.I.; De la Morena-De la Fuente, E. Agustín de Betancourt's Optical Telegraph: Geometric Modeling and Virtual Reconstruction. *Appl. Sci.* **2020**, *10*, 1857. [CrossRef]
3. Del Río-Cidoncha, G.; Rojas-Sola, J.I.; González-Cabanes, F.J. Computer-aided design and kinematic simulation of Huygens's pendulum clock. *Appl. Sci.* **2020**, *10*, 538. [CrossRef]

4. Rojas-Sola, J.I.; De la Morena-De la Fuente, E. The Hay Inclined Plane in Coalbrookdale (Shropshire, England): Analysis through computer-aided engineering. *Appl. Sci.* **2019**, *9*, 3385. [CrossRef]
5. Rojas-Sola, J.I.; De la Morena-De la Fuente, E. Agustín de Betancourt's double-acting steam engine: Analysis through computer-aided engineering. *Appl. Sci.* **2018**, *8*, 2309. [CrossRef]
6. García-Ahumada, F.; González-Gaya, C. The contribution of the Segovia mint factory to the history of manufacturing as an example of mass production in the 16th century. *Appl. Sci.* **2019**, *9*, 5439. [CrossRef]
7. Sánchez, A.; González-Gaya, C.; Zulueta, P.; Sampaio, Z.; Torre, B. Academic proposal for heritage intervention in a BIM environment for a 19th century flour factory. *Appl. Sci.* **2019**, *9*, 4134. [CrossRef]
8. Pardo Abad, C.J. Application of the digital techniques in industrial heritage areas and building efficient management models: Some case studies in Spain. *Appl. Sci.* **2019**, *9*, 4420. [CrossRef]
9. Manajitprasert, S.; Tripathi, N.K.; Arunplod, S. Three-dimensional (3D) modeling of cultural heritage site using UAV imagery: A case study of the Pagodes in Wat Maha, That, Thailand. *Appl. Sci.* **2019**, *9*, 3640. [CrossRef]
10. Saorín, J.L.; López-Chao, V.; De la Torre-Cantero, J.; Díaz-Alemán, M.D. Computer aided design to produce high-detail models through low cost digital fabrication for the conservation of aerospace heritage. *Appl. Sci.* **2019**, *9*, 2338. [CrossRef]
11. Claver, J.; García-Domínguez, A.; Sevilla, L.; Sebastián, M.A. A multi-criteria cataloging of the immovable items of industrial heritage of Andalusia. *Appl. Sci.* **2019**, *9*, 275. [CrossRef]

applied sciences

Article

Object Investigation of Industrial Heritage: The Forging and Metallurgy Shop in Taipei Railway Workshop

Yu-Hsun Chen [1], Guan-Chen Chen [2,*], Ching-Tai Wu [3], Chi-Lin Lee [4], You-Rou Chen [3], Jun-Fu Huang [3], Kuo-Hung Hsiao [3] and Jong-I Lin [3]

[1] Department of Mechanical Engineering, National Taiwan University of Science and Technology, Taipei 10607, Taiwan; yhchen@mail.ntust.edu.tw
[2] Department of Mechanical and Electro-Mechanical Engineering, Tamkang University, New Taipei City 25137, Taiwan
[3] Collections and Research Division, National Science and Technology Museum, Kaohsiung 80765, Taiwan; tiger_wu@mail.nstm.gov.tw (C.-T.W.); flyelva1022@gmail.com (Y.-R.C.); junfu@mail.nstm.gov.tw (J.-F.H.); khhsiao@mail.nstm.gov.tw (K.-H.H.); lin@mail.nstm.gov.tw (J.-I.L.)
[4] Department of History, Tamkang University, New Taipei City 25137, Taiwan; 123389@mail.tku.edu.tw
* Correspondence: gcchen@mail.tku.edu.tw; Tel.: +886-2621-5656 (ext. 2782)

Received: 28 February 2020; Accepted: 27 March 2020; Published: 1 April 2020

Abstract: As a special plant for train maintenance in northern Taiwan, the Taipei Railway Workshop was founded in 1885 and moved in 2011, reflecting the changes in Taiwan's history, transportation, and industrial technology. Now, it is planned to change the maintenance plant into a railway museum in the form of an in situ site. This study briefly introduces the historical background and present situation of the Taipei Railway Workshop and takes its forging workshop as the object for investigation and exhibition planning. According to the preservation and maintenance methods of the cultural heritage of the museum, the investigation process proposed includes four steps: Site exploration, object registration, object research, and exhibition planning. The work area in the plant is divided into shaping and forging areas, as based on the categories of the machines on the site of the forging workshop. In this study, a total of 85 industrial relics in the forging workshop are registered for systematic research. The working conditions, including machine parts for train maintenance, manufacturing processes of parts, and the relationship between in-line on-site machines and tools, of the forging workshop before closing are restored, as based on the principles of machine manufacturing, literature, and retired workers' oral histories. Finally, an in situ exhibition plan of the forging workshop is put forward based on the results of the object research.

Keywords: industrial site; railway industry; technology history; cultural heritage

1. Introduction

The Taipei Railway Workshop was a special plant for train maintenance and modification of Taiwan Railways. The Taipei Railway Workshop shows more than 100 years of Taiwan's history from 1885 to 2011, including the Qing-Ruled Period (1683–1895), the Japanese-Ruled Period (1895–1945), World War II, and the retreat of the government from the Republic of China to Taiwan (1949–now). It has status as both industrial site and cultural heritage, and is of great cultural significance to Taiwan's architecture, science, technology, historical development, labor culture, transportation, and industrial value [1,2].

The Taipei Railway Workshop is the former "Taipei Machinery Bureau", as established in 1885. In the late 19th century, Taiwan was governed by the emperors of the Qing Dynasty, and the Taipei Machinery Bureau was established under the background of the Westernization Movement to be responsible for the production of weapons and military facilities. In 1895, meaning the early Japanese-Ruled Period, the emperor of the Qing Dynasty ceded Taiwan to Japan, and the Taipei Machinery Bureau established during the Qing Dynasty was taken over by the Japanese army and changed to the "Taipei Artillery Factory", a place for weapon maintenance of the Japanese army. In 1899, with Taiwan's West Coast line railway construction plan, the General Governor of Taiwan requested that Japan's Army Ministry transfer the assets of the Taipei Artillery Factory to the Railway Department of the General Governor to become a maintenance plant for the Railway Department; therefore, the Railway Department of the General Governor of Taiwan established the "Taipei Railway Workshop" in 1900. In 1935, to respond to the growing demands for railway development in Taiwan, the Taipei Railway Workshop was relocated to its present site in Xinyi District, and became the first modern railway maintenance plant in Taiwan. From 1950 to 1970, assisted by the United States, the diesel-electric locomotive shop was newly built in the Taipei Railway Workshop to upgrade train power from steam and diesel engines to electrified power. Until 2012, when Taiwan Railways moved the maintenance base to Fugang (Taoyuan, Taiwan), the Taipei Railway Workshop was faced with the crisis of removal and resale due to the financial problems of the Railway Authority. Finally, it was preserved by the great efforts of cultural heritage groups and railway fans, and designated as a national historical site; moreover, the Ministry of Culture established the Preparatory Committee of the National Railway Museum in 2019 to carry out the core business of cultural relic collection, maintenance, research, and exhibition [1–3].

The Taipei Railway Workshop covers a total area of 193,912 m^2, including a forging workshop (forging and metallurgy shop), engine room, erecting workshop, diesel-electric locomotive shop, coach workshop, and sheet metal workshop, and is equipped with staff dormitories, employee bathhouse, and other welfare facilities for workers. The plant plan and the aerial view are shown in Figures 1 and 2, respectively [3].

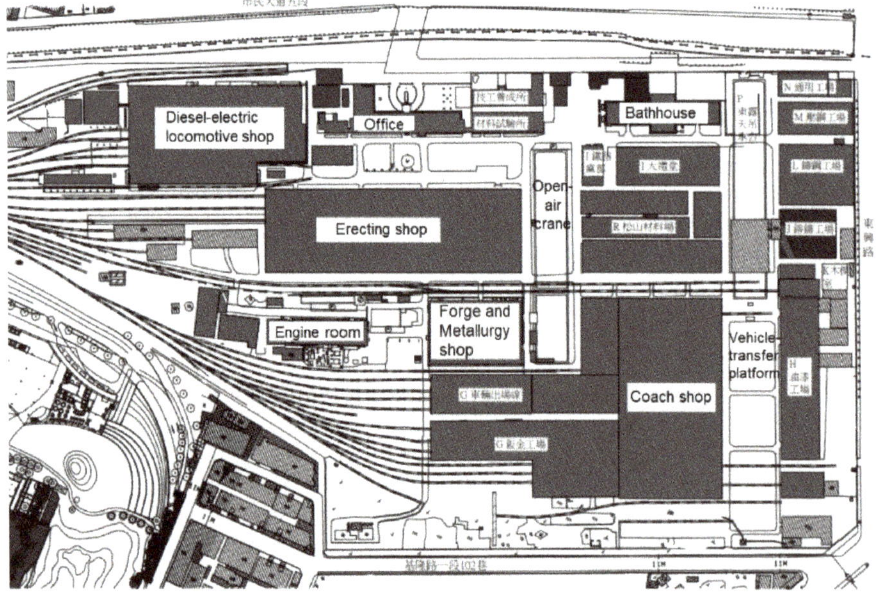

Figure 1. Plan of Taipei Railway Workshop [3].

Figure 2. View of Taipei Railway Workshop in 1930s [4].

As the Taipei Railway Workshop covers a large area, and has a large number of objects, currently, led by the Ministry of Culture, it is divided into various sections for restoration and to build an in situ museum. The purposes of this study are object research and exhibition planning of the forging workshop. Built in 1935, the forging workshop is the working field for forging and manufacturing, steel heating treatments, spring repair testing, and vehicles maintenance, thus, this area has the oldest mechanical equipment in the Taipei Railway Workshop. Figure 3 shows the working conditions of the forging workshop in the 1950s, and Figure 4 shows the present situations of the three areas of the Taipei Railway Workshop. The purposes of this study are object research and exhibition planning for the forging workshop of the National Railway Museum. Taking the forging workshop as an exhibition space, this study explains the historical background, train technology, and technology development in Taiwan driven by track vehicles.

Figure 3. Operations of the forging shop in 1950s [2].

The working area in the forging workshop can be divided into two areas, namely shaping and forging areas. For the former, spring was the main product of the shaping area. It is made of an elastic material and used to relieve shock or vibration, store energy, or measure the amount of force in mechanical devices. Since the middle of the 17th century, the leaf spring has been used on the carriage as a part of suspension [6]. To the 19th century, the leaf springs are also applied on the suspension of the trains, and it was replaced gradually by the helical spring.

Figure 4. Present situations of the plant [5]. (**a**) Sheet metal workshop, (**b**) erecting workshop, (**c**) forging workshop.

For the other area in the shop, forging is a manufacturing process in which a metal hammer is used to strike a metal workpiece to generate local compressive force to form the metal. This manufacturing technique can be traced back to thousands of years ago, and the power source of the forging devices had developed from manpower, weight [7], water [8,9], compressed air [9], to the steam and electric power.

This study conducts object inventory, organization, filing, and basic study, and interviews the workers who had been in the forging workshop. By summarizing the study results of object research and worker interviews, the complete information of file data tables of the archaeological survey is deduced and integrated, in order to convert the complex scientific knowledge into language easily understood by viewers. Meanwhile, preservation and maintenance methods are suggested, and the plans for exhibition planning, exhibition outline, promotion, and education activities are proposed [1–5].

2. Materials and Methods

With the technological evolution, aeronautical facilities, navigation equipment, and boiler units have appeared in large numbers, and due to their scientific, historical, cultural value and condition, many of them have become important cultural heritage items in museum collections. However, the preservation and maintenance of such large objects have become an urgent problem faced by museums. Many international principles and specifications have been developed for the preservation and maintenance of cultural heritage items, thus, there are many records of industrial site restoration and in situ museum planning [2,9–12].

As a part of the railway museum restoration program of the Ministry of Culture, this study organizes the cultural relics and relevant historical data of the forging workshop. Most of the cultural relics in the plant are machines, including the British 941 steam hammer, which was purchased during the Qing Dynasty in 1889 and is the most representative. In addition to sorting out the historical data of the industrial machine relics one by one, the correlation between machines, operation technology, and suggestions for cultural relic conservation are the main items of this program. Industrial machine relics equipped with electric systems can also be found, such as the machines purchased from Morita Iron Works in 1974, which are related to helical spring production. Although various industrial machine relics were purchased in different years, the display fields are set to match with them according to their functions and the processing requirements. The applications of the machines are also changed to adapt to the newly purchased machines; therefore, in addition to the considerations of the history of the cultural relics, the overall area history is another focus of this study. As shown in Figure 5, this study procedure mainly consists of four items, including site exploration, object selection, numbering and registration, object research, and exhibition planning.

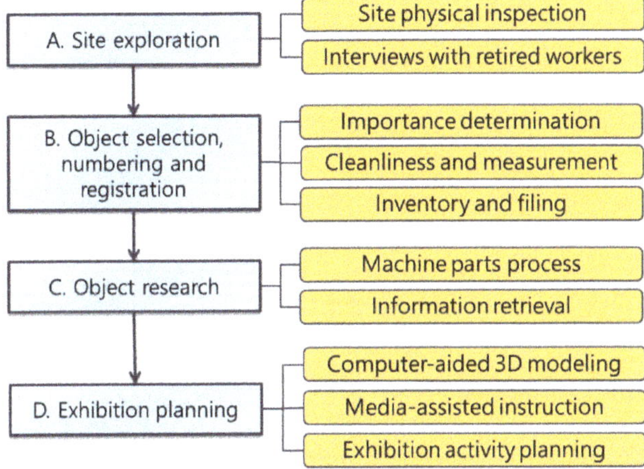

Figure 5. Study procedure.

A.   Site exploration

The members of the research team included museum researchers, cultural relic restorers, historians, and mechanics. In the first step, the people in this program conducted site explorations of the forging workshop together with the staff of the Taipei Railway Workshop (present Preparatory Office of Railway Museum) and the workers who had maintained the trains in the forging workshop, in order to confirm the current preservation situations of buildings, machines, and peripheral objects on this site.

This research team and the abovementioned staff performed the site exploration on 8 August, 2019. The workers of the forging workshop provided explanations of the site, including the parts they used to produce here, the machine operation modes, and the production processes, as well as the staff's dining and rest experiences, in order to establish the research team's basic understanding of the industrial site.

B.   Object selection, numbering, and registration

According to the basic understanding, as established in the previous step, this research team selected the objects in the forging workshop. Some on-site items were moved after the relocation of Taiwan Railways, and many fixtures and hand tools were scattered, thus, the research team focused on the manufacturing process of the mechanical components used for train maintenance, in order to select the objects, including machines in the plant, auxiliary tools, and the vehicles used to transport work pieces between machines. The object data, such as names, storage locations, external dimensions, and feature descriptions were listed for important objects, together with the numbers marked in the previous step and photos of the present situations. Furthermore, documents were created for all objects to record the detailed data of the objects, such as the units' location information, the industrial site's nature and cultural significance, investigation numbers, machine or tool labels, manufacture dates, nameplates and marks, constituents and materials, and present situation descriptions.

C.   Object research

The purpose of this step is to clarify the functions of all machines and tools during the manufacturing process. In addition to the manufacturing procedures of machine parts, the items mainly discussed are the purchase time and usage of machines, as recorded by the Taipei Railway Workshop, in order to clarify the contemporary manufacturing technologies. The machine parts made in the forging workshop were usually made of steel or alloy, as created through various steps, such as billet finishing, heating, shaping, and testing, and many machines and tools were required to complete the steps in order; for example, clamps were required to put parts into the furnace for heating; cranes or vehicles with rollers were required to convey parts to other machines for subsequent processing after heating. Hence, based on the background knowledge of the machine part manufacturing process, this step connects the relationships between all machines and tools by combining the descriptions of the workers who had worked in this field.

D.   Exhibition planning

The in situ site museum of the forging workshop was planned based on the investigation results in the first four steps, which is expected to show the historical background and cultural values of the industrial site, as well as the mechanical technology of train maintenance and the workers' life at that time. In this study, mechanical technology is transformed into easy-to-understand descriptions, and the previous production conditions in the plant are graphically explained through 3D models built with the assistance of computers. This study also planned activities to deliver the right information under the premise of considering the degree of interest by viewers. In addition, the visitors are divided into children, the general public, and students and scholars in engineering, and each group was provided with various depths of data.

3. Object Investigation

The most important part of a train's moving gear is the bogie, which carries the weight of the body and guides the train to move along the track. A bogie consists of a bogie frame, suspension system, and wheels, as shown in Figure 6. The suspension system is a linking system between the underpinning of the frame and the wheel and has the function of absorbing the kinetic energy of the vibrations, in order that riding comfort is improved [13,14].

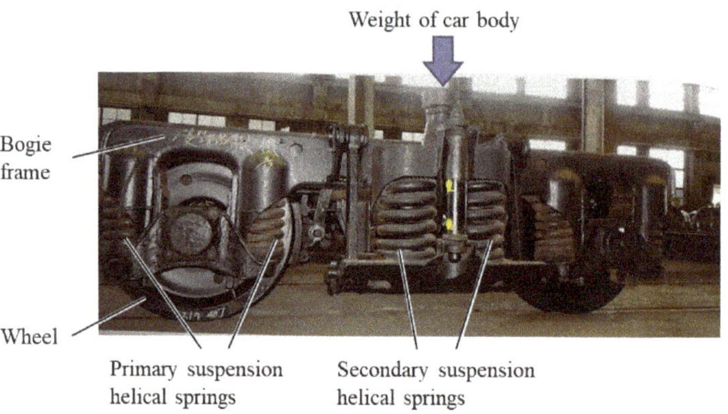

Figure 6. Composition of train bogie [13].

According to the interviewed workers of the forging workshop, leaf springs and helical springs were commonly used for the trains in early times, as shown in Figures 7 and 8. Initially, leaf springs were used for trains; however, due to the poor comfort they provided, helical springs were used instead. As entire trains had to be disassembled for overhaul, it was necessary to replace the springs if helical springs were worn or their error values exceeded the standard. Taiwan Railways used a lot of helical springs, and the manufacturing technology was not difficult; therefore, there was a series of machines used to produce helical springs at the northeast corner of the forging workshop of the Taipei Railway Workshop. After the Electric Multiple Unit (EMU) series trains were imported from the United Kingdom in 1978, almost all of Taiwan Railways' trains were converted to use air springs. On this basis, it can be inferred that the machines in the Taipei Railway Workshop are the representative production tools transferred by era and technology. At present, the steel springs seen at the Taipei Railway Workshop are only for the maintenance of early trains; for example, Chu-Kuang Express, freight trains, and Fu-Hsing Express in service now still use helical springs.

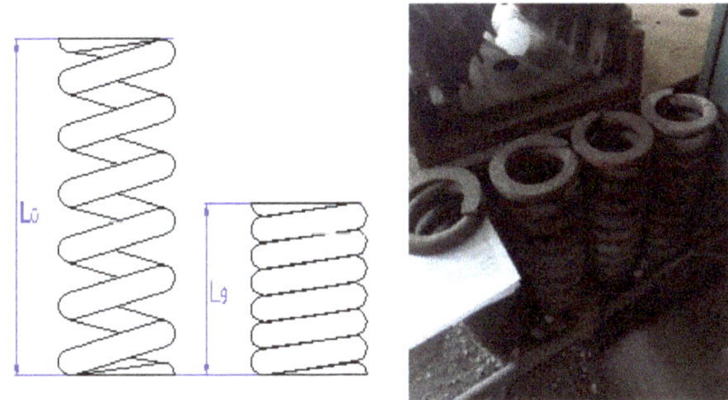

Figure 7. Helical spring.

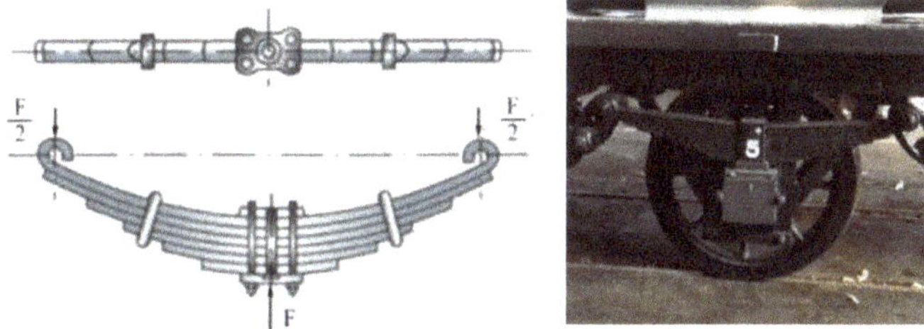

Figure 8. Leaf spring.

This study numbered and registered 85 objects, including 41 machines, 30 auxiliary vehicles and working platforms, 12 hand tools and parts, and two other cultural relics. The building of the forging workshop is 60 m long and 40.35 m wide. Based on the features of the machines and objects on site, the forging workshop is divided into three parts: "Working area", "material storage area", and "staff rest area", among which the working area is divided into the "shaping area" and "forging area", as shown in Figure 9. The shaping area is a working block where machine parts are heated, shaped, and inspected, and mainly has helical spring and leaf spring production lines according to the configurations of the tools and fields of all production lines. The main machines used in the forging area are forging hammers, which can be divided into steam-driven and electric-driven types based on the engines, including the oldest machine steam hammers used in the Taipei Railway Workshop. In addition, there are two parts in the material storage area, one is for storing the raw wire materials used to make machine parts and the other is for storing finished parts. The staff rest area has many tables, chairs, lockers, and hallstands, as well as a shrine to the patron saint of the workshop. The locations of all areas are shown in Figure 10.

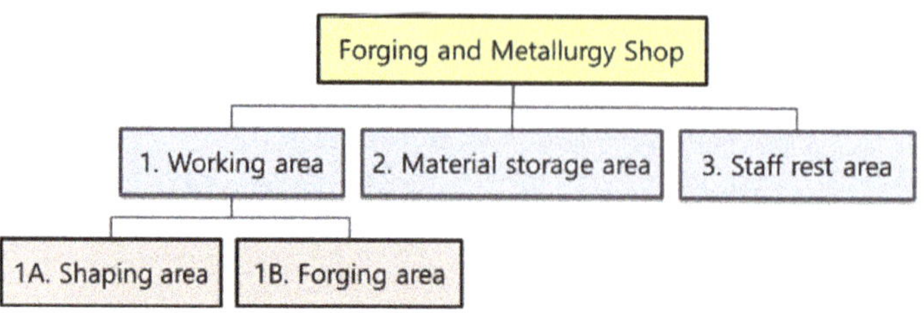

Figure 9. Classification of exhibition areas.

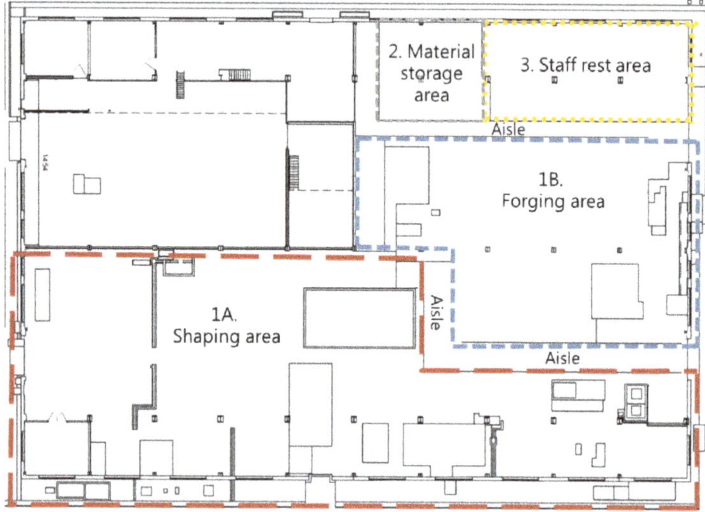

Figure 10. Locations of working blocks in the forging workshop.

3.1. Shaping Area

The shaping area is mainly used to manufacture helical springs and leaf springs, which are used for the suspension systems of trains. The manufacturing processes of these two springs, as well as the relevant machines and objects in the forging workshop, are briefly introduced, as follows.

1. Helical springs

The raw material of helical springs (Figure 7) is spring steel wire, and the manufacturing process mainly includes six steps. The relevant machines at all steps in the workshop are shown in Figure 11.

The first step is billet shearing, where the shapes of both ends of the wires used as the billets are adjusted with an inclined rolling mill (A27) after being heating in the furnace (A26). The wires are sheared to the appropriate length with a shearing machine (A22). The machines for billet shearing are shown in Figure 12.

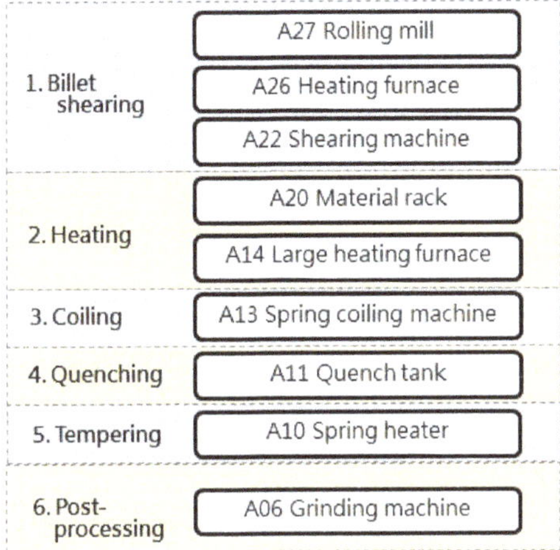

Figure 11. Flowchart of helical spring production line.

(a)   (b)

Figure 12. Billet shearing and heating. (**a**) Inclined rolling mill (A27), (**b**) shearing machine (A22), material rack (A20), large heating furnace.

The second step is heating, where the sheared steel pieces are placed on a material rack (A20), and then, put into a large heating furnace (A14). The full length of the large heating furnace is about 85 m, and the furnace temperature is about 500~800 °C. Many motor-driven rollers are equipped in the furnace to convey the billets forward to the outlet to be heated until softened.

The third step is coiling and shaping, where the required spring diameters are selected according to the appropriate spring coiling fixture (A17), in order to wind the softened wires after heating into a spiral shape through a spring coiling machine (A13). If there is any error in the spacing of the shaped springs during manufacturing, hand tools are used to make detailed adjustments. The machines for coiling and shaping are shown in Figure 13.

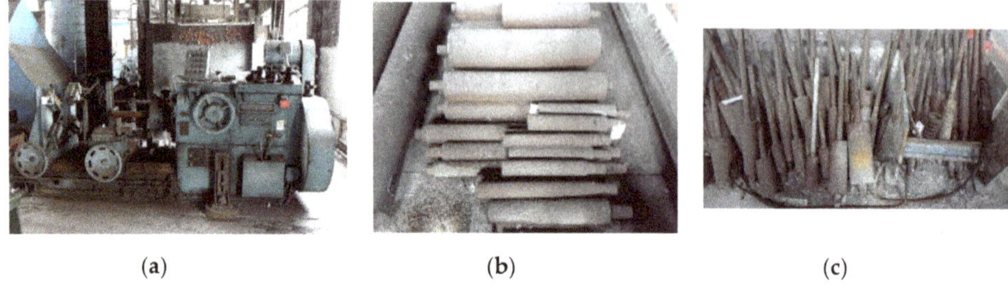

(a)   (b)   (c)

Figure 13. Coiling and shaping. (**a**) Spring coiling machine (A13), (**b**) spring coiling fixtures (A17), (**c**) spring coiling tools (A17).

The fourth step is quenching, where the shaped helical springs are immersed in a quench tank (A11) to rapidly cool the spring steels, in order to change the crystallization mode. According to the oral descriptions of the retired workers, peanut oil was once used as the quenchant for helical spring quenching in the Taipei Railway Workshop. The machines for quenching are shown in Figure 14a,b.

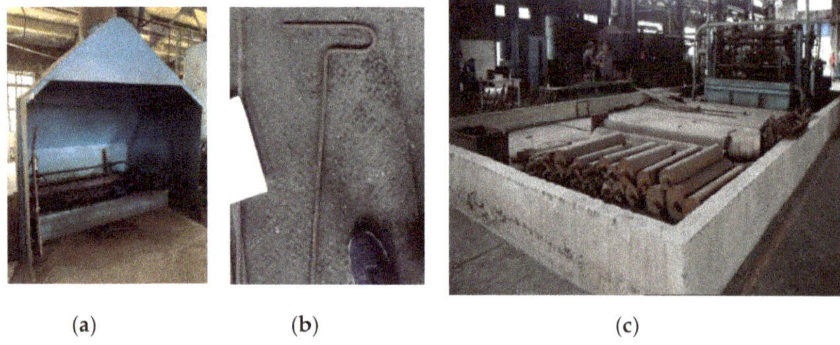

(a)   (b)   (c)

Figure 14. Quenching and tempering. (**a**) Quench tank (A15), (**b**) quench hook, (**c**) cooling tank (A11-2).

The fifth step is tempering, where the atoms of iron, carbon, and other alloying elements in steel are quickly diffused, rearranged, and recombined, in order to stabilize the steel structure. Hence, reheating the quenched helical springs in a spring heater (A10) for tempering is intended to bring the springs to an appropriate temperature below the lower critical temperature, and then, the spring ends are trimmed. The cooling tank for tempering is shown in Figure 14c.

The sixth step is post-processing, which includes grinding and rust prevention, and the finished springs are tested. Large springs are polished by an end grinding machine (A06), while small springs are directly hammered by hand, and the finished products are tested by a spring testing machine (A04). A hardness tester (A07) was also found in the field; however, as orally described by the retired workers, while this machine was originally intended for spring testing, it was found to be inapplicable. The machines for this step are shown in Figure 15.

Figure 15. Post-processing and testing. (**a**) Grinding machine (A06), (**b**) spring tester (A04), (**c**) Vickers hardness tester.

2. Leaf springs

The production process of leaf springs (Figure 8) is similar to that of helical springs, as shown in Figure 16, with the same rust prevention, testing, and other post-processing devices.

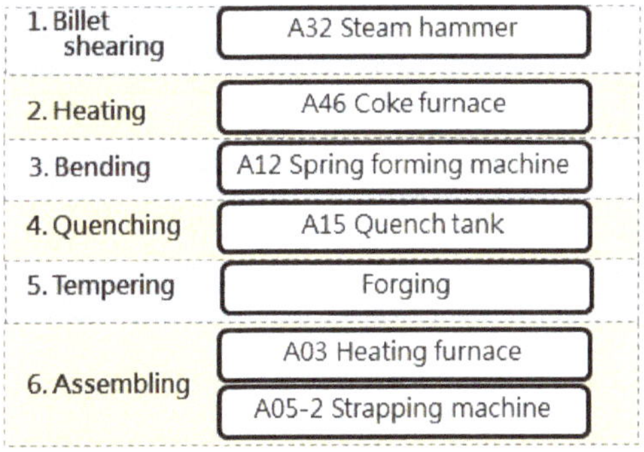

1. Billet shearing	A32 Steam hammer
2. Heating	A46 Coke furnace
3. Bending	A12 Spring forming machine
4. Quenching	A15 Quench tank
5. Tempering	Forging
6. Assembling	A03 Heating furnace
	A05-2 Strapping machine

Figure 16. Flowchart of sheet metal spring production line.

The first step is the billet shearing, where the billets are sheared to the appropriate length with a steam hammer (A32).

The second step is heating, where the sheared leaf spring billets are delivered to the coke furnace (A46) in order to heat the billets until they are soft enough to bend.

The third step is bending and shaping, where the softened billets are bent and shaped by a spring forming machine (A12). The bending length is controlled by a screw on the bender; moving the corresponding nuts by turning the screws with the handle to bend the leaf billets, as shown in Figure 17.

(a)   (b)

Figure 17. Bending, shaping, and assembling. (**a**) Spring forming machine (A12), (**b**) strapping machine (A05-2).

The fourth step is quenching, where the formed leaf springs are immersed in a quench tank (A15) to rapidly cool the spring steels.

The fifth step is tempering. The billets are forging to temper themselves.

The sixth step is assembling. A leaf spring is made up of many steel plates with different lengths, all of which are made through the above procedure, and holes are drilled in the center of all plates. Finally, the leaf spring stacks are tightened by a strapping machine (A05-2). An old photo (Figure 18a) shows the assembly of the leaf spring with a specific device. However, this device cannot be found in the forging workshop nowadays. Figure 18b shows both the helical and leaf springs made in the field in 1956.

(a)   (b)

Figure 18. Manufacture of the leaf spring (1956) (Courtesy of the Taiwan Railway Administration). (**a**) Assembly, (**b**) springs made in the forge shop.

3.2. Forging Area

Forging is a manufacturing process to shape billets into machine parts by hammer strikes. Five forging machines in different sizes still remain on the site of the forging shop in Taipei Railway Workshop, and there are many heating furnaces, cranes, pliers, and clamps in the forging area.

In the first step, the billets are clipped with clamps and put in a heating furnace, then the billets are heated to 500–800 °C, in order to soften the steel for shaping. For heavy billets, cranes are used to hang the billets, which convey billets back and forth between heating furnaces and forging machines.

The second step is forging, where parts are forged by steam hammers, forging machines, or workers with hammers, depending on their types. Figure 19 shows the use of steam hammers in the forging workshop in 1956, and it can be seen that many workers cooperated with each other for forging.

(a)   (b)

Figure 19. Forging (1956). (**a**) Two-ton steam hammer (A45) [2], (**b**) one-ton steam hammer (A32) [2].

There are five forging machines preserved in the forging shop. Inferring from No. A45 large steam hammer and No. A43 steam hammer with a large crane, they are used to process large objects. Due to low weights, No. A32 and No. A28 steam hammers are used to process small parts. Unlike the four abovementioned steam-driven machines, the other forging machine (A42) is electric-driven and used for mold forging. Figure 20 shows a set of machines for forging manufacture.

Figure 20. Current conditions of steam hammers (A45) and cranes (A47).

Among the five forging machines manufactured in Britain in 1889, the No. A32 steam hammer is the oldest machine in the Taipei Railway Workshop and has special research, preservation, and exhibition values. Figure 21 shows the words on the body of the steam hammer, including three important pieces of information. The first one is Rigby's Patent, the second is the production place in Glasgow, and the third is its production number (No. 941). According to literature, this steam hammer was manufactured by Glen & Ross LTD., a British company founded in Greenhead, Glasgow, Scotland in 1856. As a steam hammer manufacturer of considerable scale, the company had so many employees that it was able to meet the growing number of domestic orders and huge demands for export trades in the 19th century [15,16].

Figure 21. Steam hammer manufactured in 1889 (A32).

In the late 19th century, Taiwan began to introduce western machinery and made its own weapons, such as guns and ammunition, and developed its railway system. Under this background, officials at that time had sufficient motivation to buy steam hammers; however, according to the literature now collected, no direct evidence shows that the steam hammer made in 1889 came to Taiwan. Data collection and verification of this steam hammer are also the important goals of the following work.

4. Exhibition Proposal

Authors should discuss the results and how they can be interpreted in perspective of previous studies and of the working hypotheses. The findings and their implications should be discussed in the broadest context possible. Future research directions may also be highlighted.

Based on the investigation of the above objects and the studies on mechanical process technology and historical literature, this study takes the forging workshop as the research object and expects to display the items by in situ exhibition, and describes the technology of the forging workshop, its cultural significance, and the professional spirit of contemporary technicians, according to the production process of the machine parts in the forging workshop. The exhibition planning for the machines, tools, and other cultural relics on the current site mainly includes three parts: Site exhibition, description, and experience activities.

A. Site exhibition

In terms of the site exhibition, while the mechanical devices in the forging workshop have been shut down for many years, they have not been damaged, and it is possible for many machines to resume operation. The spring coiling machine (A13) and the forging hammer (A32) are representative of the machines, and are expected to perform manufacturing activities in the in situ exhibition. However, the above studies of the processes show that the helical spring billets must be heated at a high temperature of hundreds centigrade before winding. In addition to workpiece heating, the sound produced during forging is more than 100 decibels; hence, the operation mode must be adjusted to accommodate the performance in the museum. For example, if the wire billets for coiling springs are changed to tin or other soft metal, the cool coils can be directly made without heating, which can also reproduce the production process and the shapes of the finished products, although the finished products will have none of the functions of shock absorbing springs. In addition, many forging hammers are driven by steam, thus, boilers that produce steam are needed, and fuels, exhaust emissions, and other relevant problems of boilers shall be considered. However, according to the results of the object study, there is a forging machine (A28) driven by electricity, and it will be given priority if demonstration cooperation is required.

B.   Computer models and animations

Due to the limitations of machine restoration, the computer-aid techniques had been widely applied to the research and exhibitions of industrial sites, such as 3D reconstruction, animation, and virtual reality. With the help of the techniques, the visitors are able to jump out of a plane perspective and observe the full picture of the field and the details of each machine. Moreover, the analysis in the area of mechanism and materials are also helpful to realize the technology level in the ancient period, and their educational potential is clearly shown [7–12,17–21].

In this study, 3D simulation models are established with the assistance of computers, in order to record the functions, appearances, and locations of the machines and tools in the forging workshop in detail, as shown in Figure 22b,c. Moreover, in cooperation with the historical photo (Figure 19) and the present situations of the workshop (Figure 20), 3D images can also be used for visitors to understand the information of the on-site objects, such as production roles, production procedures, mechanical principles, and operation modes. In this study, the simulation models of the machines are established from the perspective of mechanical engineering, which is helpful to clarify the machines' operational principles and improve the accuracy and rationality of the models, as based on the background knowledge of the mechanical processes. The results will be helpful to the subsequent animation production, which is expected to show reasonable working methods in the workshop, including the animations of mechanical operations and parts production.

C.   Experience activities

The large machines in the workshop are dangerous in operation, thus, even if they can be restored for demonstration, it is not suitable for visitors to operate practically. Hence, this study intends to make scaled-down machines, and reduce their forces and speeds correspondingly, in order that a miniature table production line can actually work with electricity or air compressors. This activity takes the abovementioned machine parts manufacturing process and the motion relationship between the parts, as clarified by 3D models, as the reference. By operating miniature machines, visitors can experience the original process of the forging workshop to make machine parts.

With various experience activities, in addition to the function of education, the public's understanding of the Taipei Railway Workshop will be enhanced, and public attention and interest in Taipei Railway Workshop will be aroused. Moreover, the public's review rate will be improved according to the experience of various themes in different periods, and cultural heritage can be preserved and sustainably developed.

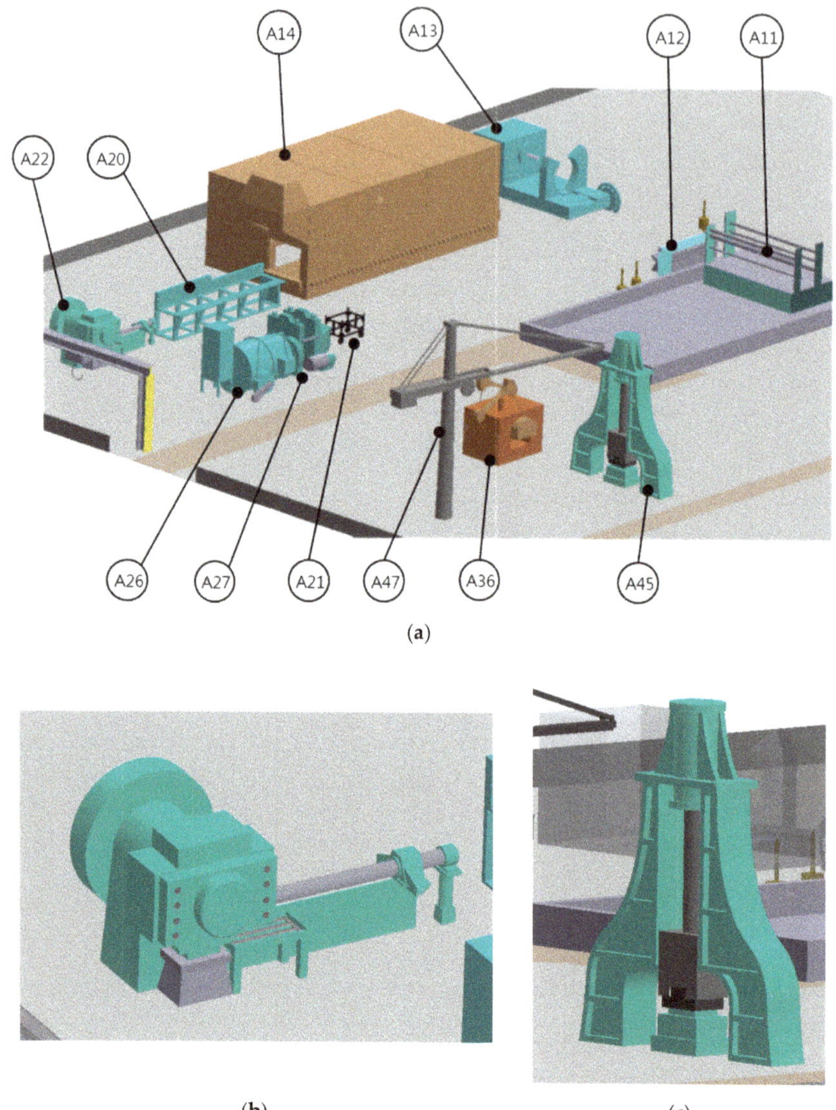

Figure 22. Computer-aided 3D models. (**a**) Part of machines in shaping area and forging area, (**b**) shearing machine (A22), (**c**) steam hammer (A45).

5. Conclusions

The Taipei Railway Workshop was in operation from 1885 to 2011, and reflects modern history, railway construction, and the mechanistic technology development processes of Taiwan, and thus, has significant cultural and technological value. After its shutdown, planning started to change the Taipei Railway Workshop from a machine plant to a railway museum; therefore, this study conducts object investigation and exhibition planning for the forging workshop.

In addition to searching literature, this study interviewed retired workers, and focused on the interviewees' working processes in the forging workshop, with particular attention paid to the manufacturing technology of train bogies and metal springs. Moreover, this study registered 85 items stored in the forging workshop, including machines, tools, and peripheral accessories. Regarding the forged parts, by considering their mechanical manufacturing procedures and cooperating with the site environment, the roles that the machines and peripheral objects played in the production of train parts are connected in accordance with the order of use. Moreover, the experiences and descriptions of site workers are integrated into the object research by referencing the interviews and survey results. In terms of object research, this study clarifies the manufacturing processes of helical springs and leaf springs in the forging workshop, as well as the site, machines, and configurations of spring manufacturing. Regarding the oldest machine, the information about the steam hammer made in 1889 coming to Taiwan is still being traced.

In situ preservation efforts are focused on preservation with the minimum intervention or permanent use. The features of the railway industry in the Taipei Railway Workshop are saved through maintaining, studying, and exhibiting the in situ museum, in order to complement the historical jigsaw of Taiwan's industrial and transportation developments. Moreover, the benefits received from exhibitions and activities will be used to maintain the museum, in order that the life of this industrial site can be sustained and cultural heritages can play their valuable roles and be sustainably managed.

Author Contributions: Data curation, C.-T.W., C.-L.L., Y.-R.C., and K.-H.H.; formal analysis, Y.-H.C., G.-C.C., C.-T.W., C.-L.L., Y.-R.C., J.-F.H., and K.-H.H.; funding acquisition, J.-I.L.; investigation, C.-T.W., C.-L.L., and K.H.H.; methodology, Y.-H.C., G.-C.C., and K.H.H.; project administration, G.-C.C., C.-L.L., K.-H.H., and J.-I.L.; resources, J.-F.H. and J.-I.L.; supervision, J.-I.L.; writing—original draft, Y.-H.C.; writing—review and editing, Y.-H.C. and G.-C.C. All authors have read and agree to the published version of the manuscript.

Funding: This work is funded by the Ministry of Culture (Taipei, Taiwan) and the Taipei Railway Workshop (Taipei, Taiwan) under the approved document, No. 1081023005, entitled "Investigation and Research on the Forging and Metallurgy Shop and the Engine Room of Taipei Railway Workshop."

Acknowledgments: The authors are grateful to the assistance provided by the Taipei Railway Workshop and Taiwan Railway Administration, especially Fugang Vehicle Depot (Taoyuan, Taiwan), Kaohsiung Railway Workshop (Kaohsiung, Taiwan), the respondents Shi-Zhou Lin, Tsan-Huang Liu, Qiu-Long Chen, Ming-Zhou Kang, Jiang-Lang Chen, and Ming-Hsun Hsieh. We would like to thank the following people who participated in this survey: The conservators Jiong-Gang Ruan and Yun-Ru Huang, the scholar Yu-Yu Huang, the assistants Bo-Hong Lin, Hao-Tian Gu, Zhen-Yu Wang, Xuan-Zhan Chen, Jung-Fang Cai, Hong-Yan Gu, Yu-Yan Li, Wan-Rong He, Chia-Ling Chen, Yen-Wen Wang, Yu Tao, Yu-Qi He, Chao-Chen Chang, Cheng-En Lee, and Hsuan-Ming Wang.

Conflicts of Interest: The authors declare no conflict of interest.

References

1. Lu, J.Z. *80 Years of the Taipei Railway Workshop: The Golden Ages of the Railway in Taiwan*; Taiwan Railway Administration: Taipei, Taiwan, 2016. (In Chinese)
2. Lo, S.S. The Research on the History and Industrial Heritage in Taipei Railway Workshop of Taiwan Railways Administration. Master's Thesis, Department of Architecture, Chung Yuan University, Taoyuan, Taiwan, 2011. (In Chinese).
3. Huang, J.M. Project Report of Restoration and Reuse plan on Taipei Railway Workshop. Available online: https://bit.ly/3ceUiVb (accessed on 22 January 2020). (In Chinese).
4. Taipei Railway Workshop. Available online: https://bit.ly/2Px0vlX (accessed on 15 February 2020).
5. Taipei Railway Workshop (National Railway Museum). Available online: https://trw.moc.gov.tw/Default (accessed on 15 February 2020).
6. Landau, D. *Leaf Springs: Their Characteristics and Methods of Specification*; Sheldon Axle Company: Wilkesbarre, PA, USA, 1912.
7. Taddei, M.; Zann, E. *Leonardo's Machines: Secrets and Inventions in the Da Vinci Codice*; Giunti Editore: Milan, Italy; Florence, Italy, 2005.
8. Hsiao, K.-H.; Yan, H.-S. *Mechanisms in Ancient Chinese Books with Illustrations*; Springer: Berlin, Germany, 2014.
9. Rojas-Sola, J.I. Study of ancient forging devices: 3D modelling and analysis using current computer methods. *Tech. Sci.* **2019**, *67*, 377–390.

10. Rojas-Sola, J.I.; De la Morena-de la Fuente, E. Digital 3D reconstruction of Agustin de Betancourt's historical heritage: The dredging machine of the port of Krondstadt. *Virtual Archaeol. Rev.* **2018**, *9*, 44–56.

11. Del Río-Cidoncha, G.; Rojas-Sola, J.I.; González-Cabanes, F.J. Computer-Aided Design and Kinematic Simulation of Huygens's Pendulum Clock. *Appl. Sci.* **2020**, *10*, 538. [CrossRef]

12. Pardo, C.J. Application of Digital Techniques in Industrial Heritage Areas and Building Efficient Management Models: Some Case Studies in Spain. *Appl. Sci.* **2019**, *9*, 4420. [CrossRef]

13. Hung, C.W. All Vehicles in Taipei Railway Administration: Illustration of the Bogies of the Passenger Trains. Self-publishing. 2019; ISBN 9789574364930.

14. Okamoto, I. Railway Technology Today 5: How Bogies Work. *Jpn. Railw. Transp. Rev.* **1998**, *18*, 52–61.

15. Carstairs, A.M. Scottish Industrial History: A Miscellany. *Econ. Hist. Rev.* **1979**, *32*, 275. [CrossRef]

16. Mayer, J. *Notices of Some of the Principal Manufactures of the West of Scotland*; The Engineering and Shipbuilding Industries: Glasgow, UK, 1876.

17. Villar-Ribera, R.; Abad, F.H.; Rojas-Sola, J.I.; Hernández-Díaz, D. Agustin de Betancourt's telegraph: Study and virtual reconstruction. *Mech. Mach. Theory* **2011**, *46*, 820–830. [CrossRef]

18. Rojas-Sola, J.I.; De la Morena-De la Fuente, E. Agustín de Betancourt's Optical Telegraph: Geometric Modeling and Virtual Reconstruction. *Appl. Sci.* **2020**, *10*, 1857. [CrossRef]

19. Rojas-Sola, J.I.; De la Morena-De la Fuente, E. The Hay Inclined Plane in Coalbrookdale (Shropshire, England): Geometric Modeling and Virtual Reconstruction. *Symmetry* **2019**, *11*, 589. [CrossRef]

20. Rojas-Sola, J.I.; De la Morena-De la Fuente, E. Geometric Modeling of the Machine for Cutting Cane and Other Aquatic Plants in Navigable Waterways by Agustín de Betancourt y Molina. *Technologies* **2018**, *6*, 23. [CrossRef]

21. Rojas-Sola, J.I.; Aguilera-García, Á. Virtual and augmented reality: Applications for the learning of technical historical heritage. *Comput. Appl. Eng. Educ.* **2018**, *26*, 1725–1733. [CrossRef]

 *applied sciences*

Article

Agustín de Betancourt's Optical Telegraph: Geometric Modeling and Virtual Reconstruction

José Ignacio Rojas-Sola [1,*] and Eduardo De la Morena-De la Fuente [2]

[1] Department of Engineering Graphics, Design and Projects, University of Jaén, Campus de las Lagunillas, s/n, 23071 Jaén, Spain

[2] Engineering Graphics and Industrial Archaeology Research Group, University of Jaén, Campus de las Lagunillas, s/n, 23071 Jaén, Spain; edumorena@gmail.com

* Correspondence: jirojas@ujaen.es; Tel.: +34-953-212452

Received: 20 February 2020; Accepted: 3 March 2020; Published: 9 March 2020

Abstract: This article shows the geometric modeling and virtual reconstruction of the optical telegraph by Agustín de Betancourt and Abraham Louis Breguet developed at the end of the 18th century. Autodesk Inventor Professional software has been used to obtain the three-dimensional (3D) model of this historical invention and its geometric documentation. The material for the research is available on the website of the Betancourt Project of the Canary Orotava Foundation for the History of Science. Thanks to the three-dimensional modeling performed, it has been possible to explain in detail both its operation and the assembly system of this invention in a coherent way. After carrying out its 3D modeling and functional analysis, it was discovered that the transmissions in the telegraph were not performed by hemp ropes but rather by metal chains with flat links, considerably reducing possible error. Similarly, it has also been found that the use of the gimbal joint facilitated the adaptability of the invention to geographical areas where there was a physical impediment to the alignment of telegraph stations. In addition, it was not now necessary for the telescope frames to be located parallel to the mast frame (frame of the indicator arrow) and therefore they could work in different planes.

Keywords: optical telegraph; Agustín de Betancourt; geometric modeling; virtual reconstruction; historical technical heritage

---

1. Introduction

The present study is part of the research line of industrial archeology whose purpose is the systematic study of the industrial memory of an era. Addressing industrial archaeology from the point of view of engineering provides a necessary vision for the correct understanding of industrial heritage, since in many occasions this study is carried out in an unscientific way, the written record of said heritage being incomplete and with the consequent risk of loss.

The work presented in this article has been developed within a research project on the work of an outstanding Spanish Enlightenment engineer, Agustín de Betancourt [1,2], analyzing his best-known inventions from an engineering graphics standpoint in order to obtain his geometric modeling [3–7].

As is well known, a first step for the recovery and study of historical technical heritage is the creation of realistic three-dimensional (3D) models. The article shows the 3D digital restitution of Betancourt's optical telegraph. This 3D model has been obtained using CAD (computer-aided design) techniques, following the objectives established in the document Principles of Seville [8] on virtual archeology which cites the London Charter [9] regarding the computer-based visualization of cultural heritage.

In 1791, the first telegraph using optical signals appeared in the work of the French Claude Chappe and his four brothers. It consisted of a dial similar to that of a watch with a moving hand that could adopt 16 positions, which each corresponded to a symbol. To indicate that the next telegraph station had received the message, it emitted an acoustic signal. However, the main drawback was in distinguishing precisely the exact position of the needle due to its size and the separation between symbols [10].

Subsequently, he developed an optical telegraph model based on five panels of the binary type (all or nothing) that assigned a code to each of the 32 possible combinations (2^5), but the results were not as expected.

Finally a third model was proposed which ignited the interest of the National Assembly of France and led to the Paris-Lille line coming into operation in 1794 along 230 km of telegraph line and 22 towers, the last one located in the dome of the Louvre, transmitting the first telegram in history [11]. Construction of the mechanisms that moved the telegraph relied on the knowledge of the Swiss watchmaker Abraham Louis Breguet, so the optical telegraph would come to be called Chappe–Breguet. The system was quite effective, although it faced the limitation that visibility between stations was not always optimal, a situation which caused numerous errors in transmission.

Simultaneously, Chappe's technological advances were widespread in numerous countries in Europe and the United States. Thus, in Sweden in 1794 Abraham Edelcrantz developed another type of telegraph that consisted of a system of 10 screens that could rotate 90° to be 'seen' or 'not seen'. Thus, the different positions of the screens formed combinations of numbers that were translated into letters, words or phrases through code books.

In the same year, 1794, the Englishman George Murray developed another telegraph, inspired by that of Edelcrantz, which consisted of six screens arranged around two columns, being able to form up to 64 possible combinations (2^6) to transmit messages [10].

In Spain the figure of the engineer Agustín de Betancourt stood out and during his stay in Paris he met Abraham Louis Breguet, with whom he could analyze in detail Chappe's telegraph. Between 1793 and 1796, Betancourt also traveled to London where he delved into Murray's telegraphic model. These visits allowed him to develop his own optical telegraph model [12], which improved on Chappe's telegraph in certain aspects.

Moreover, on 17 February 1799 King Carlos IV approved a project for the implementation of the optical telegraph in Spain, and in 1800 the first telegraph line that connected Madrid with Aranjuez began operating. However, the economic crisis that the country was going through did not allow Betancourt's invention to thrive [13].

After a review of the state of the art of the technical studies related to the Betancourt optical telegraph, only a single and interesting mechanical study has been found in the case that the telegraph stations are not aligned [14]. However, there is no study from the point of view of engineering graphics, that allows the reader to understand perfectly and in detail the operation of this historical invention that marked a milestone in the field of telecommunications. The main objective of this research is to obtain a reliable 3D CAD model of this historical invention that allows us to know in detail this outstanding engineering work. The originality and novelty of this research is that there is no existing 3D CAD model of this historical invention with this degree of detail, so it will help in an outstanding way in the detailed understanding of its operation. An educational goal is also pursued, through its exhibition on the websites of the foundations that have supported this research (Fundación Canaria Orotava de Historia de la Ciencia [15] and Fundación Agustín de Betancourt [16]), as well as in other museums of the history of technology. On the other hand, the impact of this research depends on its future uses. Among these, we can highlight:

- Performing a static analysis with CAE (computer-aided engineering) techniques in order to determine whether the invention was well dimensioned and supported the demands of its operation.
- Developing applications of virtual reality and augmented reality to promote the interaction of the user with the model that will help them to better understand its operation and appreciate for each element or system its denomination and the materials from which it was manufactured.
- Incorporating WebGL models into a website.

- Printing in 3D using additive manufacturing together with an animation created by a photorealistic organizer for its exhibition in a museum, interpretation center, or foundation.

The remainder of the paper is structured as follows: Section 2 presents the materials and methods used in this investigation. Section 3 includes the main results of the process of geometric modeling and discussion of them in order to explain the operation of this device, and Section 4 states the main conclusions.

2. Materials and Methods

The material used in this research was obtained from two files on the Betancourt Digital Project website of the Canary Orotava Foundation for the History of Science [17], assigned for its digitalization by the National School of Bridges and Roads of ParisTech University. The first of these is manuscript 826 (MS 826), consisting of 3 sheets (Figures 1–3), and a descriptive memory of 26 pages written by Betancourt and Breguet [18] and divided into two parts: one dedicated to the explanation of the 3 sheets and the operation of the optical telegraph, and a second where the authors defend their ideas on how telegraphic language should be. The second is the 1806 (MS 1806) which contains several documents related to the optical telegraph [19].

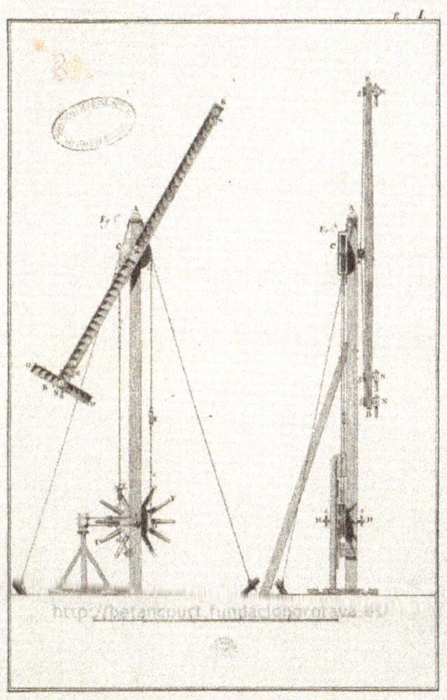

Figure 1. Elevation (**left**) and profile (**right**) views of the optical telegraph by Agustín de Betancourt and Abraham Louis Breguet [18] (Courtesy of Fundación Canaria Orotava de Historia de la Ciencia).

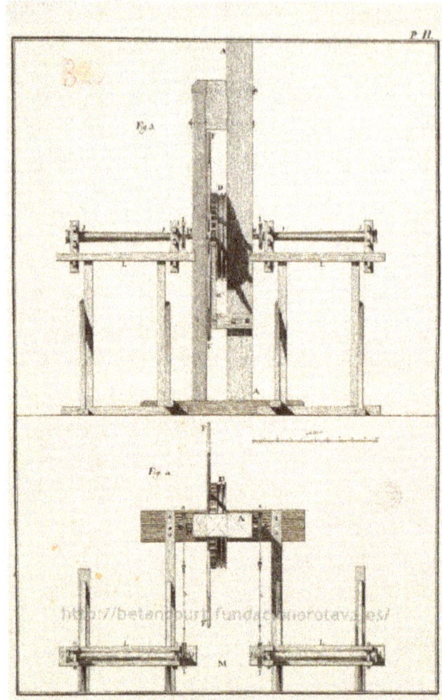

Figure 2. Longitudinal (**top**) and cross (**bottom**) sections of the optical telegraph by Agustín de Betancourt and Abraham Louis Breguet [18] (Courtesy of Fundación Canaria Orotava de Historia de la Ciencia).

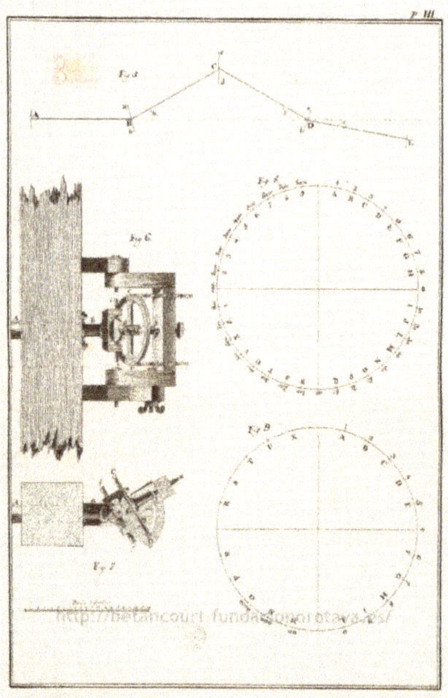

Figure 3. Observation angles and signal system of the optical telegraph by Agustín de Betancourt and Abraham Louis Breguet [18] (Courtesy of Fundación Canaria Orotava de Historia de la Ciencia).

The drawings used to model the telegraph appear with their respective graphic scales, which has greatly facilitated the correct dimensioning of each of the telegraph elements. The drawings, despite being plans as they are conceived today, are rich in constructive detail and surprise with their clarity. The possible doubts that have been generated, in some cases, by the lack of orthographic views are resolved thanks to the memory written by Betancourt, which details the operation and purpose of each element.

The first sheet (Figure 1) presents a general view of the elevation and the profile of the optical telegraph. In some parts, it omits details of the frame or support, which can hide important parts of the mechanism. The second sheet (Figure 2) shows a longitudinal and a cross section that highlight the relationship between the frames, showing in detail the transmission from one to the other by means of the transmission that joins the pulleys. The third sheet (Figure 3) is divided into two parts: the left part shows the elevation and plan views of the gimbal joint, while the right part illustrates the positions that the optical telegraph can take. The upper right part shows how the optical telegraph can be placed on a different work plane from the immediate telegraph stations (previous and rear), while the lower right part details the signs transmitted by the optical telegraph according to the direction the indicator arrow adopts.

Thanks to these drawings it has not been necessary to make assumptions about the geometry of the elements to be modeled, a rare aspect in other inventions of the same author, since the function of the plans at the time was to describe the function of the invention and not the to give details regarding its construction.

The research methodology has followed these steps:

1. Transcription and translation of the descriptive reports of the two files described above (MS 826 and MS 1806).
2. Proportional printing of the three sheets of the MS 826 file in an A3 format to directly take the measurements of the elements and apply the graphic scale present in the drawings.
3. Identification of the different elements and their functionality in the set.
4. Determination of the geometry and dimensions of each element, once its dihedral projections have been identified thanks to our knowledge of descriptive geometry [4].
5. Obtaining the 3D model of each element using parametric CAD software.
6. Assembly of all elements of the set by applying restrictions and joints. The restrictions (dimensional, geometric and of movement) allow us to define the degrees of freedom that the adjacent elements must have in their movement, while the joints do not present degrees of freedom. All this has allowed us to obtain a 3D CAD model of the set that is coherent and functional, reflecting the mechanical and structural characteristics of the mechanism.

The entire process described above is shown as a summative flowchart in Figure 4.

For CAD modeling tasks the software used has been Autodesk Inventor Professional [20] which has allowed us to obtain the 3D CAD models of each element, generating a file with extension (.ipt) and finally the assembly of the whole, generating a file with extension (.iam). Moreover, each element has been assigned material with certain physical properties in order to obtain a model that is closest to reality.

Three-dimensional CAD modeling techniques constitute a very important tool in the process of the study and design of historical heritage, for example, in the fields of cultural heritage [21], aerospace [22], civil [23], horological [24], architectural [25], and more generally, in the study of any virtual model [26–28], as a previous step to CAE analysis [29,30].

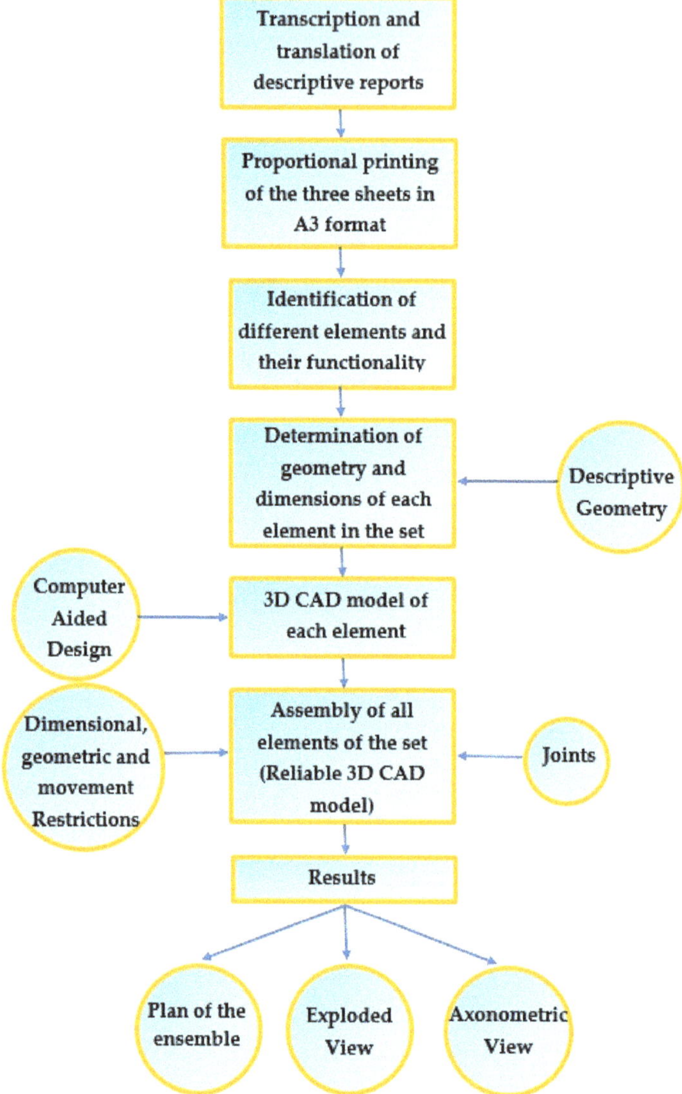

Figure 4. Summative flowchart of the methodology followed in the 3D modeling process.

3. Results

3.1. Considerations and Functioning

In order to give the reader a complete idea about the analysis of the invention of Agustín de Betancourt it is necessary to explain its operation. Figure 5 shows a plan of the ensemble with an indicative list of the different elements that form it which will serve to illustrate the operation of the mechanism, and Figure 6 shows an exploded view of the overall invention for a better understanding of the direction and order of assembly of the different parts of the optical telegraph.

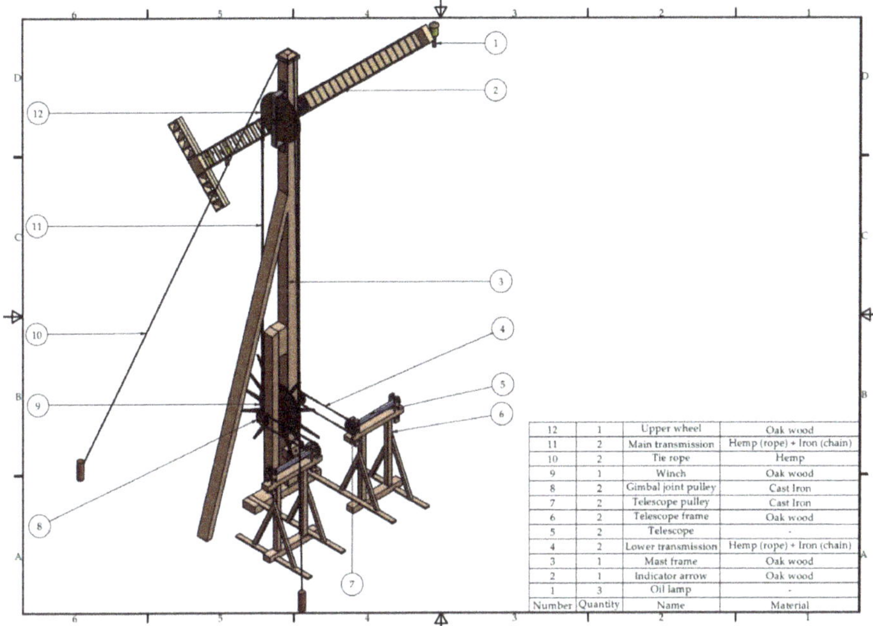

Number	Quantity	Name	Material
12	1	Upper wheel	Oak wood
11	2	Main transmission	Hemp (rope) + Iron (chain)
10	2	Tie rope	Hemp
9	1	Winch	Oak wood
8	2	Gimbal joint pulley	Cast Iron
7	2	Telescope pulley	Cast Iron
6	2	Telescope frame	Oak wood
5	2	Telescope	-
4	2	Lower transmission	Hemp (rope) + Iron (chain)
3	1	Mast frame	Oak wood
2	1	Indicator arrow	Oak wood
1	3	Oil lamp	-

Figure 5. Plan of the ensemble of the optical telegraph with an indicative list of all its elements and materials.

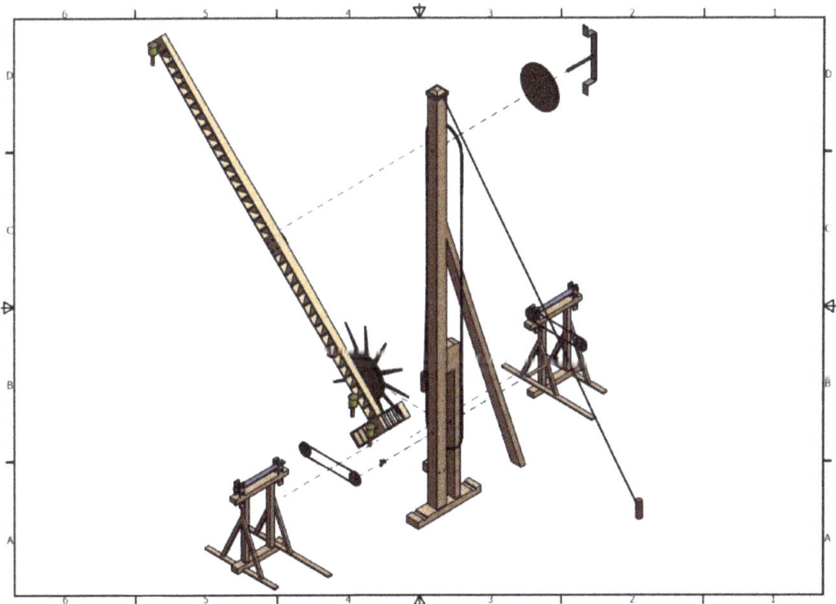

Figure 6. Exploded view of the 3D CAD model.

The operation of the Betancourt optical telegraph is based on the relationship between the indicator arrow (2) and the mast (3) with which it forms a certain angle easily visible over large distances. The directions that the arrow can adopt are taken over preset positions corresponding to the transmission code signs. The indicator arrow revolves around an axle that allows it to adopt any direction of an imaginary circle, dividing it into thirty-six sectors which correspond to numbers or letters of the alphabet. Comprising 36 divisions, each sector covers an angle of 10° sexagesimal: the first positions correspond to letters of the alphabet and the rest to ten figures (from 0 to 9).

Firstly, in order to change the position of the indicator arrow a manually operated main transmission (11) is necessary. In the lower part of the mast, there is a winch (9) with 36 slots, which was operated by the operator until the tip of said indicator arrow was placed on the slot with the mark of the alphanumeric character to be transmitted. In a similar way, the main transmission (11) communicates the movement of the winch (9) to the upper wheel (12) which is integrated with the axle of the indicating arrow so that turning the winch rotates the arrow, adopting a certain position. The aforementioned main transmission (11) is singular, since it is formed by two long pieces of hemp rope fastened at its ends by a chain of flat links joined by bolts, very similar to those used in current transmissions. Finally, the telegraph operator looks through the telescope's eyepiece (5) for the arrow position of the preceding telegraph to make sure that its arrow has the same position.

Each telegraph has two telescopes perfectly fixed in two wooden frames (6), which gives them stability and serves to keep them pointing in a certain direction (the one of the nearest telegraph: before or after). This telescope can rotate on its longitudinal axle without ever losing the direction of the immediate telegraph, and in addition it is housed in a pulley (7) arranged transversely to its main axle. This pulley (7) is also joined by a transmission to another pulley (8) joined in turn by a gimbal joint to the axle of the winch. This gimbal joint allows the telescopes to be rotated at the same time as the winch, thanks to a lower transmission (4) existing between them, similar to the main transmission (11) mentioned above.

Furthermore, the telescope's eyepiece is provided with a meridian wire that serves to show the position of the indicator arrow of the telegraph itself. Thus, if the operator observes that the indicator arrow of the preceding telegraph changes the winch will rotate until the position of its arrow is the same as that of the preceding telegraph. In the same way, the operator can see if the rear telegraph station indicates the position that he has transmitted and will not change position until it is transmitted correctly.

An important consideration is that telegraph stations do not have to be perfectly aligned for proper operation, that is to say that it is not necessary that the frames be arranged in parallel, since the gimbal joint allows the transmission work plane to be different from that of the circle formed by the indicator arrow.

Finally, the indicator arrow has three oil lamps (1) that allow the telegraph to continue operating at night, since they indicate the direction of the arrow in the dark, the operating mode being the same as in full daylight.

3.2. Optical Telegraph Assembly Process

The Betancourt and Breguet optical telegraph was a simpler mechanism to assembly than any of the contemporary telegraphs, although in any case it is a structure that when its mast and arrow are aligned has a minimum total height of 9.85 m (if it is true that, depending on the needs of the location, it could reach more than twenty meters high). A precision mechanism of these dimensions requires a fairly careful assembly, and therefore the study of the mechanism and its modeling using CAD techniques has enabled the authors of the article to gain a special awareness of the delicate assembly of the telegraph which we will now describe.

In the first place, for the correct assembly of the telegraph, it is necessary to judiciously choose the location of the telegraph station. The operator should have a direct and clear view of the immediate stations. Thus, a location aligned as much as possible with respect to the preceding and subsequent station was chosen, facilitating installation and reducing the possible optical error. Once this location was chosen the plane of the telegraph (plane defined by the circumference described by the indicator arrow) was located as parallel as possible to the planes of the immediate telegraphs, and if this was impossible to measure (there were already compasses at the time), taking the direction perpendicular to the one drawn by the segment that joined the immediate telegraphs. Thus, the frame that held the mast (3) should be perfectly aligned knowing these data, and the structure of the mast should not be raised until a series of elements were placed on it, since once it was lifted it, would be necessary to place a large structure to assemble the rest of the elements.

With the mast on the ground, the first element that was to be placed was the winch (9) with its axis, positioning it perpendicular to the structure of the mast. Once in position the shaft had to be fixed by placing the two gimbal joints on the outer faces of the frame, which do not prevent the shaft from rotating but do prevent its axial movement.

The next operation was to place the upper wheel (12) with its own axle. In order to do this, the axle had to be placed in its exact location, the wheel housed and finally the metal structure fixed which gave a second point of support to the axle. Thus, the wheel and the winch were still in place to insert the main transmission (11) before lifting the mast. The main transmission was one of the most delicate elements to place in the mechanism. As explained in the description of the invention, the main transmission was formed of a middle zone and two extreme zones. The middle zone was constituted by a hemp rope and the extreme zones were two chains of flat links, and the join of both elements was achieved by means of a tensor that when turned moved its ends closer together. For the assembly of the main transmission one of the tensors needed to be detached from the two ends. Thus, the part of the chain transmission was placed both in the groove of the wheel (12) and in the groove of the winch (9) and once in position the tensor was put in place and the main transmission tensioned (11). This operation had to be performed at least twice since the transmission was double, that is to say there were two main transmissions in parallel. Once it had been verified that, when turning the winch, the upper wheel turned in the same direction, it was time to lift the structure.

In order to raise and fix the structure two ropes (10) were necessary, acting as struts of about 12 m in length, which served to give stability to the structure and prevent its swaying. One end of the rope was tied to the head of the mast while the other was attached to an anchor in the ground. Moreover, the structure of the mast has a counterweight on the side opposite the arrow that also supports the mast, which also helps to gain stability in the structure. Finally, once the structure was lifted, it was fixed to the ground by two long anchor bolts.

After this operation the frames (6) of the telescopes were installed. This was one of the most delicate tasks since the frame had to be fixed to the ground once the telescope was pointing in the right direction, something that was also achieved by means of anchor bolts, and in order to obtain the correct direction it was essential to have a signal indicating at least the position of the immediate telegraphs. Thus, once the two frames were oriented, the telescopes were placed and fixed (5) by means of a wooden clamp which allowed them to rotate but not to move axially. The telescopes on the other hand were not normal since the end of the telescope's eyepiece was housed in a pulley (7). Once the telescope was placed and after making sure that it pointed clearly to the nearest station, it was rotated until the meridian wire in the eyepiece was arranged vertically.

The transmission mechanism between the telescope pulley and the winch made it easy to control the position of the telescopes from the winch. The gimbal joint was connected by a pin to a pulley (8), and this through a lower transmission (4) communicated the movement to the telescope pulley (7). The assembly process, therefore, was very similar to that of the main transmission: first the gimbal joint pulley (8) was mounted by the aforementioned pin, and then the transmission was placed on them, inserting the chain into the grooves perimeter of both pulleys. The transmission also had a central part of hemp rope and two extreme zones of flat metal links. In the assembly process, it was very important that both the telescope and the winch were fixed in the same position (the meridian wire in vertical position and the winch pointing to the vertical position of the indicator arrow). Again, one of the two link chain tensors had to be detached from it while the lower transmission was placed in its position and tightened until it was tight. This operation had to be performed twice, once for each telescope.

The final step to finishing the assembly was the placement of the indicator arrow (2) on its axle. Obviously this process was simple if the mast measured only a few meters, but if the mast measured more than 10 m the process presented more complexity, since it was necessary to install the structure using a crane and with the help of many operators, because the indicator arrow alone measured 6.60 m. The indicator arrow mounted separately with its three oil lamps (1) and its transverse base should be placed on its axle in an upright position and with the horizontal arm in the position closest to the ground. Once inserted it should be fixed to the horizontal axle by four bolts in order to secure it perfectly. This horizontal axle was undoubtedly the piece that was going to endure the majority of the stresses to which the mechanism would be subjected, and the useful life of the mechanism depended on its correct installation. After this step, the mechanism would be in perfect condition for use.

3.3. Three-Dimensional Modeling of the Parts and Final Assembly of the Three-Dimensional CAD Model

The modeling process of the optical telegraph is long and should be performed in great detail, paying close attention to the detailed information offered by the drawings.

First, the mast frame is modeled (Figure 7). This wooden structure is the skeleton of the invention, and it is important to respect the distances of the holes where the different axes will be housed. The mast is 7.45 m high and rests on a support that is 1.32 m long. This structure is not enough to give stability to the mast, therefore, from the central body of the mast, there is a support that allows the base to be extended to 2.27 m. In a similar manner, Betancourt, aware of the wind resistance that the structure presents, proposed a system of supports to guarantee the stability of the structure, and on the other hand the proposed dimensions of the mast and its structure will always depend on the topographic conditions of the environment. Its function is structural, but it must also be able to place the indicator arrow at a sufficient height, and if the conditions require it to be extended.

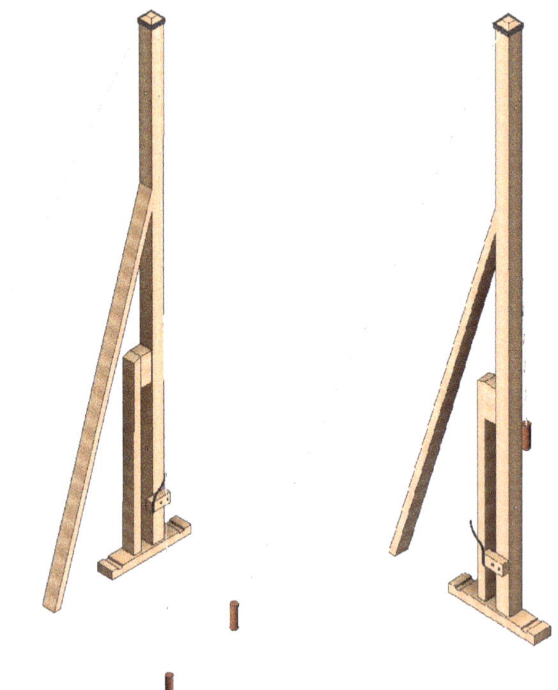

Figure 7. Mast frame (posterior and anterior view).

The next structure to model is the upper wheel (Figure 8). This is a wooden pulley of 75 cm in diameter in whose perimeter there are two grooves where the main transmission chain will be housed. This wheel is fixed to the mast thanks to a metal axle, and this in turn is supported externally by a metal structure that is screwed to the mast. The structure allows the axle to be perfectly horizontal and the wheel to rotate freely.

Figure 8. Detail of the upper wheel.

On the other hand, there is the wooden winch, which is one of the most important elements of the mechanism (Figure 9). This is a wooden wheel of 75 cm in diameter in which there are also two grooves in the perimeter zone, as in the upper wheel, but not in the middle area of the wheel rather towards one side. The other part of the wheel is drilled every 10° so that the sectors in which the wheel is divided can be perfectly distinguished. These divisions will be used to position the indicator arrow in a certain direction. The winch in addition has 12 arms of 1 m in length that facilitate its maneuverability, and in turn is crossed by a metal shaft that has a double function, positioning the winch in the mast support and allowing it to rotate freely.

Figure 9. Detail of the winch.

Figure 10 shows the position of the winch on the mast frame, giving a very clarifying overall view.

Figure 10. Detail of the position of the winch on the mast frame.

The main transmission, the one that connects the upper wheel and the winch (Figures 11–13), is the most complex modeling element of the mechanism. If the drawings and Betancourt's memory are taken carefully, it is discovered that the transmission that communicates the movement from one to the other is not achieved by a simple hemp rope, since the use of a rope presents the inconvenience of its sliding inside the slit due to the inertia of the movement of the indicator arrow. In order to avoid this slippage, Betancourt proposes a mixed rope-chain transmission system. The belt-like transmission has a central part of the rope and a part of the metal chain that comes into contact with the pulleys. Between them, there is a metal tensor whose mission is to provide greater tension to the transmission or to facilitate its repair tasks.

The modeling of the hemp rope is simple, although from the structural point of view it is very problematic. Most CAD software represent ropes as static elements when their behavior is dynamic. Also, the section and the length of the rope change as a tension is applied to it and no conventional design software is able to simulate this. Therefore, from an aesthetic point of view, the design of the rope does not offer much complication, but if it is intended to use the CAD model for a subsequent CAE analysis, then the model is insufficient.

The modeling of the chain of flat links is very laborious since it is necessary to model each link with its own bolt, defining a relationship between them that allows it to rotate with respect to each other. As can be seen in Figure 14, this chain of links in its description is very similar to that of a current flat link chain transmission, although evidently Betancourt did not know this type of transmission and defines links much larger than those of a current chain.

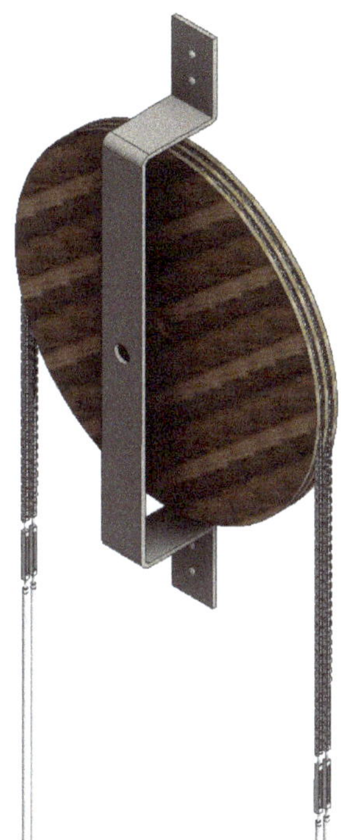

Figure 11. Detail of the upper wheel transmission.

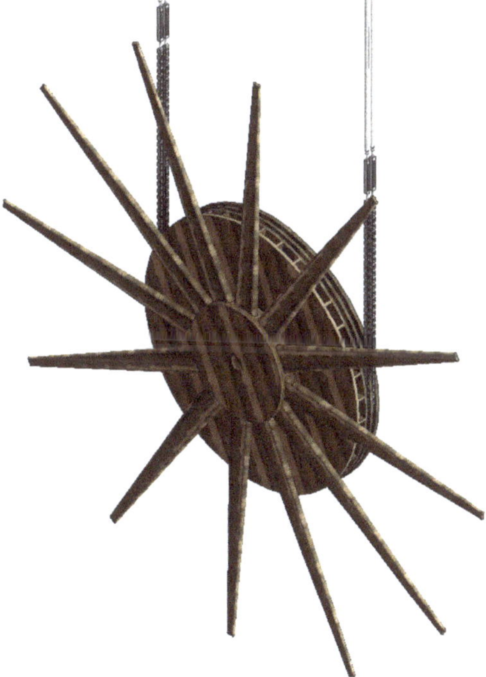

Figure 12. Detail of the transmission on the winch.

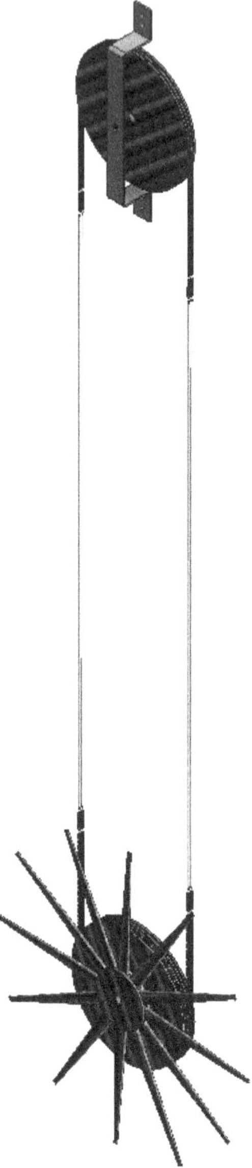

Figure 13. Detail of the transmission between the upper wheel and the winch.

It is known that at the time there were no standardized elements and that each metallic element had to be handmade, so a chain of these characteristics was very complex and required a great manufacturing effort. Therefore, in practice, and for a large number of telegraph stations, it is not surprising that rope transmissions were used instead of mixed transmissions. Finally, it should be noted that the transmission was double in order to ensure the correct synchronization between the indicator arrow and winch.

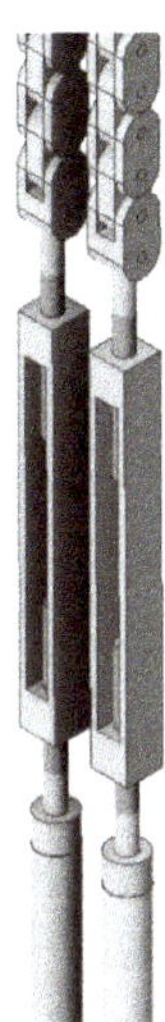

Figure 14. Detail of the chain of links and tensors of the transmission.

Next the telescope frame is modeled (Figure 15). This frame is a simple wooden structure 1.53 m high, 1.45 m wide and 1 m long, which has the function of housing the telescope through which the nearest telegraph station is observed. For this the wooden clamp must at all times allow the telescope to rotate freely while preventing its longitudinal movement, although the telescopes frames do not have to be aligned with respect to the mast frame. In this article, they are presented as aligned, but their position depends on the direction in which the nearest station is located.

Figure 15. Telescope frame.

Once the telescope has been modeled, a 22 cm diameter metal pulley is attached to the rear of the telescope where the eyepiece is located (Figure 16). When the pulley rotates, the telescope rotates in unison.

Figure 16. Detail of the placement of the telescope pulley on its frame.

The next step in the modeling, also technically complex, is that of the pulley attached to the gimbal joint (Figure 17). The first element to model in this case is the end of the winch axle where a U-shaped metal piece with holes drilled in its two tips is to be placed. The next element is a circular piece that has two main axes, each axle being a metal rod of circular section topped by a square section that prevents it from passing through the hole. In addition, one of the axes goes through the two holes of the U-shaped metal tip mentioned above. It is important to define the relationship of these two elements correctly so as to allow the circular piece to rotate freely.

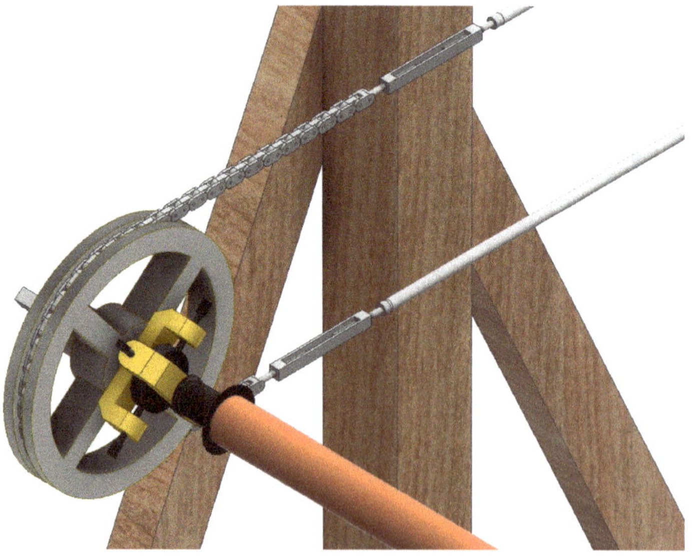

Figure 17. Detail of the gimbal joint with its pulley.

Next a metal pulley of dimensions and characteristics similar to that used in the telescope is modeled, but with a difference: located on one of its faces it has a U-shaped metal piece similar to that of the axle but placed perpendicularly and facing its ends. The tips of this piece are also drilled and its purpose is to house the free shaft that the round metal part possessed.

This articulation, made with these two U-shaped metal parts and the two-axle circular piece, is what is commonly called gimbal joint or universal articulation, and it allows the rotational movement of one axle to be converted into the rotational movement of another axis arranged in an unaligned direction. The use of this joint is what allows the telescope frame to not necessarily have to be aligned with the mast frame.

The transmission that is used between the metal pulleys (the one attached to the gimbal joint and the telescope pulley), has similar characteristics to the transmission between the upper wheel and the winch, but with a lower number of links (Figure 18).

Figure 18. Detail of the transmission between metal pulleys.

After modeling the transmission, we can see how moving the winch makes both telescopes rotate simultaneously at the same angle (Figure 19).

Figure 20 shows the modeled indicator arrow. As can be seen, it is a simple T shaped structure of 6.60 m in length, with its interior emptied so that it offers less wind resistance. The indicator end is signaled by an oil lamp, which could be lit if necessary, and the transverse part of the arrow (opposite end) is in turn signaled by two oil lamps. In this way, one end was also distinguishable from the other in the dark.

Figure 19. Turning ratio between the winch and the telescope pulley.

Figure 20. Indicator arrow.

Once the indicator arrow has been modeled, it must be placed in its screwed position by means of four screws to the upper wheel (Figure 21) so that when the operator manipulates the winch the arrow rotates in unison with the upper wheel.

Figure 21. Detail of the placement of the indicator arrow on the mast.

Figure 22 shows an axonometric view of the optical telegraph assembled with all its elements.

3.4. Discussion and Findings

The optical telegraph presented here possesses a series of characteristics that made it the most advantageous telegraph of its time.

First, it is a telegraph with a relatively simple assembly. The mechanism used for its operation is very basic, although somewhat more complex than that of the Edelcrantz panels, but in comparison, the set of transmissions and telescopes greatly simplifies its handling, since only one operator is needed who does not have to know the signs he transmits.

Figure 22. Optical telegraph assembled with all its elements.

Secondly, the transmission method using only an indicator arrow makes the message clearly gain compared to the method of panels or two arms used by other telegraphs. The use of an indicator arrow allows the operator to use 36 positions, allowing him to use the conventional language of signs and numbers and not a complex network of codes.

Thirdly, the indicator arrow allowed the use of oil lamps to distinguish the signals at night, so as long as the visibility was adequate transmission of the message was not reduced to the hours of daylight.

On the other hand, the optical telegraph also presented a series of drawbacks that weighed down the life of the telegraph and that finally led it to disappear shortly after it was put into operation. Interestingly, among these drawbacks, only one is of a technical nature, and it was precisely the one that would go unnoticed at the time.

The use of telescopes entailed a small error that was not possible to avoid at the time. When sighting a telegraph at certain angles with a telescope (it has already been seen that it is not easy to accurately locate both the immediate telegraph plane and the observation plane), some optical deformation occurs in the reading. For example, for a 30° angle difference, the observation error is 4.6°. The measuring instruments of the time, especially compasses, did not allow for more precision when aligning these planes, although a 30° offset would be an excessive assumption since the sectors for the same signal are 10°. Therefore, it cannot be assured, that in any of the messages no mistake was made, although in general with that maximum error it would be perfectly acceptable.

On the other hand, the use of the gimbal joint could also generate some error in the transmission. In addition, at the time there were any mechanical studies on the movement of said gimbal joint, which is not homogeneous and, therefore, the transmission was not a synchronized movement between pulley (8) and winch (9). A study of the limits of the telegraph [14] provides more complete information on the subject, although the conclusion is that the error was negligible, partly due to the use of two gimbal joints for two telescopes.

The last drawback presented by the optical telegraph was the candle effect. A slender structure located in a geographical location with adverse atmospheric situations was subject to great stresses, although so were the structures that supported the rest of the telegraphs. To avoid this effect, Betancourt devised the indicator arrow as a lightened structure, that is, instead of being a solid wooden pole, it consisted of a wooden body, empty inside, but consistent thanks to the slats that reinforced its structure. This arrow was less resistant to wind and reduced stresses in the support structure.

However, the main problem faced by the telegraph was that it was a patented invention (Betancourt–Breguet), and the conditions of the patent (Breguet had the exclusive rights to its installation) were viewed suspiciously and judged by political interests. The installation of the telegraph was not a small cost, since it took a good number of telegraph stations to cover the routes. In Spain while the influence of Betancourt remained the optical telegraph project went ahead. Thus originally a first test line, Madrid-Cádiz, was to be implemented, although only the first phase of the Madrid-Aranjuez line was installed since the disagreements with Manuel Godoy eventually condemned the project. In France, the payment of a patent to a foreign company was the main problem they faced. From a technical point of view, Betancourt and Breguet repeatedly demonstrated that their invention had obvious advantages over the French Chappe telegraph, but Chappe's better personal contacts gave his invention the final advantage.

When modeling the invention in 3D, it was discovered that the transmissions of the optical telegraph were not made of rope, although the models that exist of the invention in some parts of the world have not taken this aspect into account.

This observation is not an insignificant matter since if Betancourt and Breguet had proposed a telegraph with a rope transmission it would have been necessary to calibrate it every so often. The mass of the arrow with its axle is 83 kg and therefore is not excessive as it is a lightened structure, but its resistance to the wind and the arrow's own inertia when moving would have led to the hemp rope sliding on the pulley, causing a not unsubstantial error which would need to be corrected from time to time.

Betancourt does not propose a chain transmission arbitrarily, but does so with a very clear intention. The use of a metal chain inserted into a wooden pulley (or in a wooden winch) and the tensor of this chain made the transmission translate its geometry onto the wood which acted as a mold. In this way there was no risk of the transmission sliding through the groove of the pulleys, so a correction that was costly to solve and that periodically required a large adjustment time was avoided.

Finally, the gimbal joint is another of the great contributions of the invention which demonstrates the great mastery of mechanics of the Spanish engineer. Its use facilitated the adaptability of the invention to geographical areas where there is a physical impediment for telegraph stations to be aligned, and furthermore, when using this articulation it was not necessary for the telescope frames to be parallel to the frame of the indicator arrow and they could therefore work on different planes.

4. Conclusions

The optical telegraph of Agustín de Betancourt and Abraham Louis Breguet is one of the most important legacies in the history of engineering. Not surprisingly, it was the first telegraph that operated in the Iberian Peninsula and it was one of the first in the world to operate.

This article has had as its main objective to obtain 3D modeling and virtual reconstruction of this invention respecting its original project to the maximum detail, using the Autodesk Inventor Professional software. Thanks to this 3D modeling, it has been possible to explain in detail both its operation and the assembly system of this invention in a coherent way, allowing us to discover the virtues of the invention and the advantages it presented from the point of view of engineering over its contemporary telegraphs.

The methodology employed in this research has been based on our knowledge of descriptive geometry and the use of direct empirical measurement techniques on the three drawings of the invention file which presented a graphic scale. Thus, after modeling each element of the set with the abovementioned parametric design and engineering software, it was necessary to establish restrictions (dimensional, geometric, and of movement) as well as of joints in order to obtain a coherent and fully functional 3D CAD model.

From this 3D CAD model it has been possible to obtain detailed geometric documentation of each element, as well as an axonometric view, a plan of the ensemble with an indicative list of all the elements and their material, and an exploded view with indication of each element in the assembly, which has made it very easy to fully understand the detailed operation of the assembly and the function of each element within it.

Among the main discoveries, it has been found that the transmissions in the telegraph were not performed by hemp ropes rather by metal chains with flat links, since the error introduced was much smaller than with the use of ropes. Similarly, it has also been found that the use of the gimbal joint facilitated the adaptability of the invention to geographical areas where there was a physical impediment for the telegraph stations to be aligned and, in addition, facilitated the non-obligation of the telescope frames to be parallel to the frame of the indicator arrow, thus enabling them to work in different planes.

This research methodology can be applied to a multitude of technical historical heritage inventions studied over the centuries and will allow us to rescue these notable contributions of technology to society, highlighting their legacy thanks to the techniques of modeling and virtual reconstruction as a first step towards a CAE study.

Author Contributions: Formal analysis, E.D.l.M.-D.l.F.; funding acquisition, J.I.R.-S.; investigation, J.I.R.-S. and E.D.l.M.-D.l.F.; methodology, J.I.R.-S. and E.D.l.M.-D.l.F.; project administration, J.I.R.-S.; supervision, J.I.R.-S.; validation, E.D.l.M.-D.l.F.; visualization, J.I.R.-S.; writing—original, draft, J.I.R.-S. and E.D.l.M.-D.l.F.; writing—review and editing, J.I.R.-S. and E.D.l.M.-D.l.F. All authors have read and agreed to the published version of the manuscript.

Funding: This research was supported by the Spanish Ministry of Economic Affairs and Competitiveness under the Spanish Plan of Scientific and Technical Research and Innovation (2013–2016), and the European Fund for Regional Development (EFRD) under grant number [HAR2015-63503-P].

Acknowledgments: We are very grateful to the Fundación Canaria Orotava de Historia de la Ciencia for permission to use the material of Project Betancourt available at their website. Similarly, the authors of the article express their sincere thanks to Pedro Jesús Pancorbo Rico for his work with the initial CAD model used in this research. Also, we sincerely appreciate the work of the reviewers of this article.

Conflicts of Interest: The authors declare no conflicts of interest.

References

1. Muñoz Bravo, J. *Biografía Cronológica de Don Agustín de Betancourt y Molina en el 250 Aniversario de su Nacimiento*; Acciona Infraestructuras: Murcia, Spain, 2008. (In Spanish)
2. Martín Medina, A. *Agustín de Betancourt y Molina*; Dykinson: Madrid, Spain, 2006. (In Spanish)
3. Rojas-Sola, J.I.; De la Morena-de la Fuente, E. The hay inclined plane in Coalbrookdale (Shropshire, England): Geometric modeling and virtual reconstruction. *Symmetry* **2019**, *11*, 589. [CrossRef]
4. Rojas-Sola, J.I.; Galán-Moral, B.; De la Morena-de la Fuente, E. Agustín de Betancourt's double-acting steam engine: Geometric modeling and virtual reconstruction. *Symmetry* **2018**, *10*, 351. [CrossRef]
5. Rojas-Sola, J.I.; De la Morena-de la Fuente, E. Digital 3D reconstruction of Agustin de Betancourt's historical heritage: The dredging machine of the port of Krondstadt. *Virtual Archaeol. Rev.* **2018**, *9*, 44–56. [CrossRef]
6. Rojas-Sola, J.I.; De la Morena-de la Fuente, E. Geometric Modeling of the Machine for Cutting Cane and Other Aquatic Plants in Navigable Waterways by Agustin de Betancourt y Molina. *Technologies* **2018**, *6*, 23. [CrossRef]
7. Rojas-Sola, J.I.; De la Morena-de la Fuente, E. Agustin de Betancourt's wind machine for draining marshy ground: Approach to its geometric modeling with Autodesk Inventor Professional. *Technologies* **2017**, *5*, 2. [CrossRef]
8. Principles of Seville. Available online: http://smartheritage.com/wp-content/uploads/2016/06/PRINCIPIOS-DE-SEVILLA.pdf (accessed on 20 February 2020).
9. London Charter. Available online: http://www.londoncharter.org (accessed on 20 February 2020).
10. Villar-Ribera, R.A. Estudio del Telégrafo de Agustín de Betancourt. Ph.D. Thesis, Technical University of Catalonia, Tarrasa (Barcelona), Spain, 15 February 2013.
11. Bahamonde, A.; Martínez, G.; Enrique Otero, L. *Las Telecomunicaciones en España. Del Telégrafo Óptico a la Sociedad de la Información*; Ministerio de Ciencia y Tecnología: Madrid, Spain, 2002. (In Spanish)
12. Museo Virtual CEDEX. Available online: http://www.museovirtual.cedex.es/detalle-maqueta-cedex.html?id=140 (accessed on 20 February 2020).
13. Olivé Roig, S. *Historia de la Telegrafía Óptica en España*; Ministerio de Transporte, Turismo y Comunicaciones: Madrid, Spain, 2010. (In Spanish)
14. Villar-Ribera, R.; Hernández-Abad, F.; Rojas-Sola, J.I.; Hernández-Díaz, D. Agustin de Betancourt's telegraph: Study and virtual reconstruction. *Mech. Mach. Theory* **2011**, *46*, 820–830. [CrossRef]
15. Fundación Canaria Orotava de Historia de la Ciencia. Available online: http://fundacionorotava.org (accessed on 20 February 2020).
16. Fundación Agustín de Betancourt. Available online: http://www.fundacionabetancourt.org (accessed on 20 February 2020).
17. Proyecto Digital Betancourt. Available online: http://fundacionorotava.es/betancourt/ (accessed on 20 February 2020).
18. Betancourt, A.; Breguet, A. Mémoire sur un Nouveau Télégraphe et Quelques Ideés sur la Langue Télégraphique. Available online: http://fundacionorotava.es/pynakes/lise/bregu_teleg_fr_01_1797/ (accessed on 20 February 2020).
19. AA.VV. Description du Télégraphe Inventé par les Citoyens Bréguet et Betancourt. Available online: http://fundacionorotava.es/pynakes/lise/vario_teleg_fr_01_1796 (accessed on 20 February 2020).
20. Shih, R.H. *Parametric Modeling with Autodesk Inventor 2016*; SDC Publications: Mission, KS, USA, 2015; ISBN 978-1-63057-030-9.
21. Ullah, A.M.M.; Watanabe, M.; Kubo, A. Analytical point-cloud based geometric modeling for additive manufacturing and its application to cultural heritage preservation. *Appl. Sci.* **2018**, *8*, 656. [CrossRef]
22. Saorín, J.L.; Lopez-Chao, V.; de la Torre-Cantero, J.; Díaz-Alemán, M.D. Computer aided design to produce high-detail models through low cost digital fabrication for the conservation of aerospace heritage. *Appl. Sci.* **2019**, *9*, 2338. [CrossRef]
23. Rojas-Sola, J.I.; De la Morena-de la Fuente, E. The hay inclined plane in Coalbrookdale (Shropshire, England): Analysis through computer-aided engineering. *Appl. Sci.* **2019**, *9*, 3385. [CrossRef]
24. Del Río-Cidoncha, G.; Rojas-Sola, J.I.; González-Cabanes, F.J. Computer-aided design and kinematic simulation of Huygens's pendulum clock. *Appl. Sci.* **2020**, *10*, 538. [CrossRef]

25. Manajitprasert, S.; Tripathi, N.K.; Arunplod, S. Three-dimensional (3D) modeling of cultural heritage site using UAV imagery: A case study of the pagodas in Wat Maha That, Thailand. *Appl. Sci.* **2019**, *9*, 3640. [CrossRef]

26. Kasparova, M.; Halamova, S.; Dostalova, T.; Prochazka, A. Intra-oral 3D scanning for the digital evaluation of dental arch parameters. *Appl. Sci.* **2018**, *8*, 1838. [CrossRef]

27. McEwen, I.; Cooper, D.E.; Warnett, J.; Kourra, N.; Williams, M.A.; Gibbons, G.J. Design & manufacturing of a high-performance bicycle crank by additive manufacturing. *Appl. Sci.* **2018**, *8*, 1360. [CrossRef]

28. Koca, G.O.; Bal, C.; Korkmaz, D.; Bingol, M.C.; Ay, M.; Akpolat, Z.H.; Yetkin, S. Three-dimensional modeling of a robotic fish based on real carp locomotion. *Appl. Sci.* **2018**, *8*, 180. [CrossRef]

29. Rojas-Sola, J.I.; De la Morena-De la Fuente, E. Agustin de Betancourt's double-acting steam engine: Analysis through computer-aided engineering. *Appl. Sci.* **2018**, *8*, 2309. [CrossRef]

30. Rojas-Sola, J.I.; De la Morena-De la Fuente, E. Agustin de Betancourt's mechanical dredger in the port of Kronstadt: Analysis through computer-aided engineering. *Appl. Sci.* **2018**, *8*, 1338. [CrossRef]

 *applied sciences*

Article

Computer-Aided Design and Kinematic Simulation of Huygens's Pendulum Clock

Gloria Del Río-Cidoncha [1], José Ignacio Rojas-Sola [2,*] and Francisco Javier González-Cabanes [3]

[1]   Department of Engineering Graphics, University of Seville, 41092 Seville, Spain; cidoncha@us.es
[2]   Department of Engineering Graphics, Design, and Projects, University of Jaen, 23071 Jaen, Spain
[3]   University of Seville, 41092 Seville, Spain; fj.gonzalezcabanes@gmail.com
*   Correspondence: jirojas@ujaen.es; Tel.: +34-953-212452

Received: 25 November 2019; Accepted: 9 January 2020; Published: 10 January 2020

Abstract: This article presents both the three-dimensional modelling of the isochronous pendulum clock and the simulation of its movement, as designed by the Dutch physicist, mathematician, and astronomer Christiaan Huygens, and published in 1673. This invention was chosen for this research not only due to the major technological advance that it represented as the first reliable meter of time, but also for its historical interest, since this timepiece embodied the theory of pendular movement enunciated by Huygens, which remains in force today. This 3D modelling is based on the information provided in the only plan of assembly found as an illustration in the book *Horologium Oscillatorium*, whereby each of its pieces has been sized and modelled, its final assembly has been carried out, and its operation has been correctly verified by means of CATIA V5 software. Likewise, the kinematic simulation of the pendulum has been carried out, following the approximation of the string by a simple chain of seven links as a composite pendulum. The results have demonstrated the exactitude of the clock.

Keywords: Huygens's pendulum clock; computer-aided design; virtual recreation; kinematic simulation; CATIA V5

---

1. Introduction

The topic of this article is the subject of numerous publications [1–3], which demonstrates the interest therein and follows the line of research on the recovery of technical historical heritage [4–9], in particular that of antique clocks [10]. The sun and water were, in ancient times, the first references to the measurement of time. After these types of clocks, perfected by Egyptians and Greeks, hourglasses appeared, and subsequently, clocks with cog mechanisms were invented in the 13th century. In the 15th century, the first spring-powered timepieces were developed in Germany [11,12].

In the 16th century, major geographical discoveries provoked an interest in the precise determination of longitude at sea. Therefore, the lunar clock was invented, as were astronomic clocks [13,14]. By the 17th century, the cog-regulating pendulum of Christiaan Huygens had appeared, as had the discoveries of Robert Hooke with respect to the law of elasticity; these mechanical improvements contributed towards the invention of the first balance-spring-regulated pocket chronometer from Hooke and Huygens, which constituted a milestone in precision horology.

In the 18th century, John Harrison solved the problem of longitude, and certain elements were perfected, such as the profile of the teeth of the cogs and the compensation of the pendulum; these aspects contributed towards improving accuracy. Finally, in the 19th century, with the industrial revolution, mass production and mechanization appeared. These improvements in manufacturing technologies constituted major contributions towards the improvement of the precision of watchmaking mechanisms.

Christiaan Huygens (The Hague, Netherlands, 1629–1695) was one of the central characters of the scientific revolution in the 17th century alongside Francis Bacon, René Descartes, Galileo Galilei, and Isaac Newton. With training in physics, mathematics, and astronomy, he contributed notably by perfecting Kepler's telescope, thereby discovering Titan and the Orion Nebula. In 1656, he invented the pendulum clock, after adding a pendulum to a clock driven by weights, so that the clock would keep the pendulum moving and regulate its progress [15]. After this discovery, he stated the pendulum theory employed to measure the acceleration of gravity and its variations with altitude and latitude [16–18].

He subsequently developed, with little success, a portable chronometer to make it easier for seafarers to determine geographical longitude at sea, based on the double suspension of the pendulum between cycloidal blades. Interest in his theories is still shown in various studies [19,20].

Huygens published several books, among which *Horologium* [21,22] stands out, in which he defines the first pendulum clock, and *Horologium Oscillatorium* [23–25], his masterpiece, where he describes the isochronous pendulum clock and studies the properties of the cycloid and of geometric curves in general.

Huygens's pendulum clock is a notable example of technical historical heritage, although there are many projects for the recovery of cultural heritage in general. A good sample of these include the mechanical lion of Leonardo Da Vinci [26], the initiative of the Canarian Orotava Foundation of History of Science to study the machines invented by the distinguished Spanish engineer Agustín de Betancourt [27], the virtual reconstruction of the device by Juanelo Turriano to raise water from the Tagus river to the city of Toledo, Spain [28], the clock of the Cathedral of Santiago de Compostela, Spain [29], the Church of Santa María de Melque in Toledo, Spain [30], and the Church of San Agustín de la Laguna in Tenerife, Spain [31].

The objective pursued in this historical investigation consists of obtaining a reliable three-dimensional model and the simulation of its movement through the parametric software CATIA V5 [32] of the isochronous pendulum clock, as designed by Christiaan Huygens and published in 1673, to verify its correct operation and underline its exactitude. Its selection is due to the significant technological advance that it represented as the first reliable time meter, especially thanks to the use of the properties of the cycloid, and also for its historical interest, since the whole theory of pendulum movement that Huygens enunciated remains in force.

The contribution of this research lies in the fact that there is currently no 3D CAD model of this historical invention with any degree of detail, which would help in the comprehension of its general operation, which is anything but simple. This operation has been verified through simulations from its virtual recreation, generating videos to check the properties of the cycloid (isochronous movement), and therefore, the accuracy of its functioning.

Another objective of this 3D model is educational, as it is designed for display in a History of Technology museum or interpretation centre. The impact of this research depends on the future uses of the model, which include:

- Development of applications of virtual reality and augmented reality to promote the interaction of the user with the model which will help users to better understand its operation and to appreciate its classification for each element or system.
- Incorporation of the WebGL model into a website.
- 3D printing using additive manufacturing together with an animation created by a photorealistic organizer for its exhibition in a Museum, Interpretation Centre or Foundation.

2. The Isochronous Pendulum Clock

As previously stated, in 1673, Huygens published his masterpiece *Horologium Oscillatorium*, which was divided into five parts, wherein he describes the isochronous pendulum clock and studies the properties of the cycloid and geometric curves in general. He also indicated a certain length of the pendulum to produce a certain number of movements, and observed that the pendulum that marked seconds should measure approximately 99.42 cm.

This research has been carried out based on the first three parts of the book. The first describes the clock (Figure 1) as consisting of three main subsets: The gear train, the mechanism of transformation of the circular movement into an oscillatory movement, and the elements associated with the movement of the pendulum.

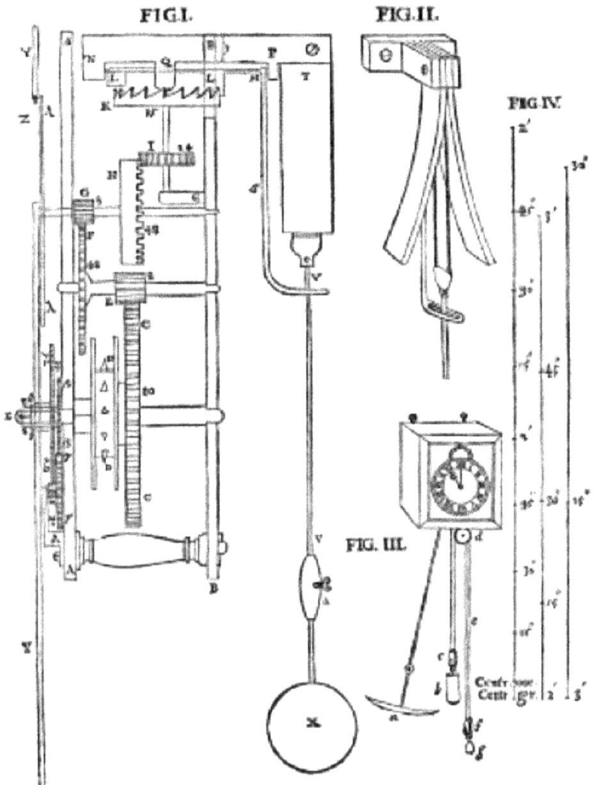

Figure 1. Isochronous pendulum clock of Christiaan Huygens according to *Horologium Oscillatorium*.

The gear train is configured on three axes interconnected by wheel-pinion assemblies that are responsible for managing the clock hands. The mechanism of the transformation of the circular movement into an oscillatory movement is composed of a crown wheel, K, that rotates around the vertical axle and transforms its rotation into oscillation of the rod, S, thanks to a horizontal axle with vanes, LM, held by the so-called 'Gnomons' (parts where the axles are housed). Finally, the elements associated with the pendulum are the pendulum itself, its support, and the cycloidal blades that limit its movement on either side, transmitted thanks to the rod, S.

In the second and third parts of the work, the theoretical foundations are enunciated and demonstrated to achieve the isochrony of the pendulum, by forcing the pendulum to move tangent to two cycloidal blades, which describe its extremes, the cycloid, and, therefore, a tautochronous trajectory (isochronous movement). In this way, the pendulum achieves a constant period (one second) independent of the amplitude of the movement.

The understanding of the operation of the mechanism was not easy, although the main challenges faced in this research include the creation of the designs of the gears of different tooth profiles, the design of the two cycloidal blades (two coupled escapement-driven blades), the simulation of the pendulum movement, and especially, the behaviour of the strings that support the bob of the pendulum.

3. Sizing, Design, and Assembly of the Elements of the Isochronous Pendulum Clock

The clock design has been carried out independently for each of the three component parts: The gear train; the pendulum with the cycloidal blades; and the clamp (Gnomon P), the dimensions of which are explained in detail in the *Horologium Oscillatorium*. For this modelling phase, the Sketcher and Part Design modules of CATIA V5 were applied. These 3D CAD modelling techniques provide a fundamental tool in the process of designing of technical historical heritage, for example, in aerospace heritage [33], as a previous stage of CAE (computer-aided engineering) analysis [34,35] and, in general, as an application for any example of cultural heritage [36].

The only information available regarding the operation of the pendulum clock appears as a brief description of the component parts: The number of teeth of each wheel, the speed that each axle must have, and the dimensions that the pendulum must have in order to mark seconds. However, there is no information available regarding the shape and dimension of each piece, and hence the design proposed in this research has been made only with the graphic information available in Figure 1.

It is well-known that pendulum clocks are characterized by using an oscillating weight to measure time accurately, and that they enjoy the advantage that the pendulum behaves like a harmonic oscillator, that is, its oscillation cycles are produced in equal time intervals (periods), and only depend on its length (isochrony). For this reason, these pendulum clocks must remain in a fixed position, since any displacement would affect the movement of the pendulum and the accuracy of its operation. Therefore, the sizing of a pendulum clock is completely parametric, and depends only on the length of the pendulum, which uniquely implies a gear ratio.

The pendulum clock can therefore be considered as being divided into two distinct parts: on the one hand, there is gear train and, on the other hand, the pendulum which is made up of the cycloidal blades and the clamp (Gnomon P), the sizing of which is explained in detail in the book.

3.1. Gear Train

The gear train design is undoubtedly the first step in the redesign phase of the clock: It is decisive in the size of the resulting clock, since the distance between the axles is directly related to their primitive diameter. The gear train can be differentiated from the rest of the gears because the speed ratio of each pair of gears that are geared is directly related to the ratio of the diameters of the wheels, and this, in turn, is related to the ratio of the number of teeth. Therefore, the different speeds of rotation of each axle will be the speeds of the hands of the clock, which depends on the relation of the number of teeth and not of the size.

To obtain the speed ratios between gears, it suffices to use wheels whose number of teeth are in an inverse relationship, and the size of these gears will be determined by their own modulus (the gearing wheels have the same modulus).

Given the absence of detailed information regarding the dimensions of the elements, it has been assumed that the gear train as a whole would measure vertically one third or, at most, half the length of the pendulum. This assumption has enabled us to estimate that the primitive diameter, d, of the largest wheel (C—crown wheel), which has 80 teeth, is 240 mm, for which the modulus, m (the quotient between the primitive diameter and the number of teeth), is 3. It is known that the modulus is a standardized parameter, since it is decisive in the construction and calculation of gears, in such a way that all the data of the gearwheels is expressed as a function of said parameter. Thus, the complete pendulum clock design is based on this unit of measure.

Once all the gearing wheels were defined, the distance between the axles could be determined as the semi-sum of the primitive diameters of said wheels. Additionally, the decision was taken that the gear pressure angle should be the same, with the objective of transmitting movements and not forces and, on the other hand, considering that the distance between the AA and BB plates that form the structure of the clock (Figure 1) is 150 mm and the dimensions are 200 mm wide by 552 mm high.

Since the transmission between the gears must be carried out continuously to allow uniform movements, any of several curves could be chosen for the design of the tooth profile: The cycloid, the epicycloid, the hypocycloid, and the involute of the circle [37]. Finally, the profile of the involute of the circle was chosen because it can be traced with arcs of circumference, thereby obviating the freehand trace, and with it the error introduced [38].

To this end, our method of choice was that of Grant's odontograph [39], which results in a curve that is a very good approximation to the theoretical profile. The design parameters of wheels and sprockets were chosen while taking into account both the requirements of the chosen tracing method and the basic conditions of the wheels and sprockets that engage [39,40].

Likewise, the 3D modelling of all teeth followed a common pattern of operation: First, a sketch was made, in which the tooth profile was drawn; secondly, an extrusion (pad) of that sketch was performed, to subsequently produce a circular matrix (circular pattern) of the extrusion; finally, a hole was made to accommodate the corresponding axle.

Once all the gears were defined, the positions of each of the five axles of the mechanism were determined by means of the semi-sum of the primitive diameters of each gear. This design was completely satisfactory, since it was verified that there were interferences between the gears by sliding the teeth between them, and allowing the precise operation of the clock.

Due to the complexity of the final model (consisting of 49 different components), and to the fact that presenting the assembly plan would be extremely extensive, as would the detailed drawings of each dimensioned piece and the exploded view of the assembly to show its order of assembly in the search for the reproducibility of the work, we opted to include Figure 2, which shows only the front view of the 3D CAD model and indicates the nine wheels (I, H, G, F, E, C, B, Y, and P) and the pinion, S, as well as other main parts of the assembly. Likewise, Table 1 shows the main dimensional values determined in this study for all nine wheels and the pinion.

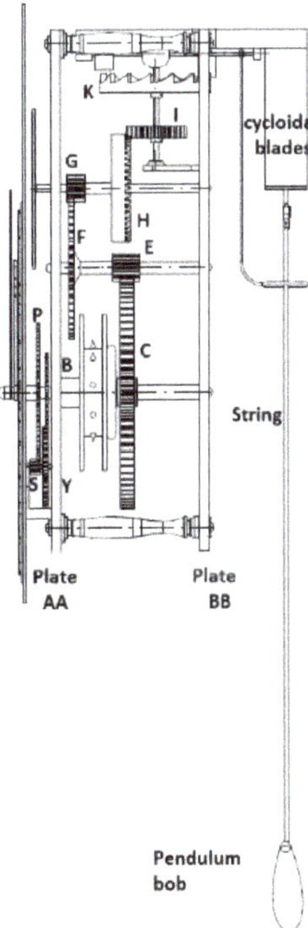

Figure 2. Front view of the 3D CAD model with an indication of all wheels and the pinion, as well as the main parts.

Table 1. Main dimensional values determined for all wheels and pinion.

Wheels	Number of Teeth	Primitive Diameter (m)	Modulus	Pressure Angle (°)
S (Pinion)	6	0.012	2	26
G & E	8	0.024	3	23
I	24	0.060	2.5	14.5
Y & B	30	0.078	2.6	14.5
H	48	0.120	2.5	14.5
F	48	0.144	3	23
P	72	0.144	2	26
C (Crown wheel)	80	0.240	3	23

3.2. Cycloidal Blades

These elements constitute the key to the success of this pendulum clock. The cycloid is traced by faithfully following the method explained by Christiaan Huygens in his book *Horologium Oscillatorium*, where it is defined as the cyclic curve that is generated by a point when rolling (without sliding) a circle along a line (Figure 3).

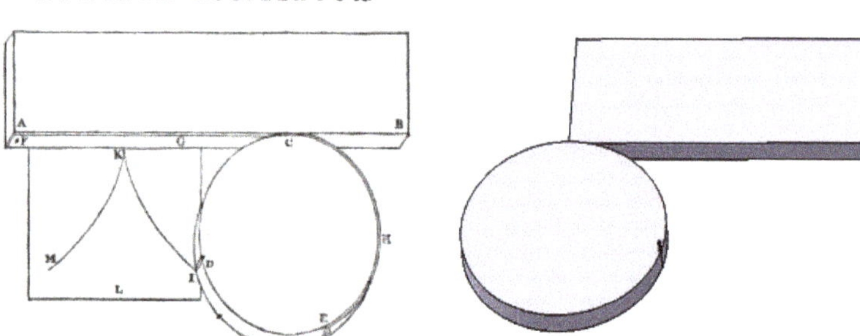

Figure 3. Tracing the cycloid with CATIA V5 according to *Horologium Oscillatorium* and a 3D CAD model.

This curve can be obtained in the DMU Kinematics Simulator module of CATIA V5, in which the software is instructed to draw the trajectory of a previously defined point of the circle, whereby this cycloidal curve describes a succession of points joined by a spline (Figure 4).

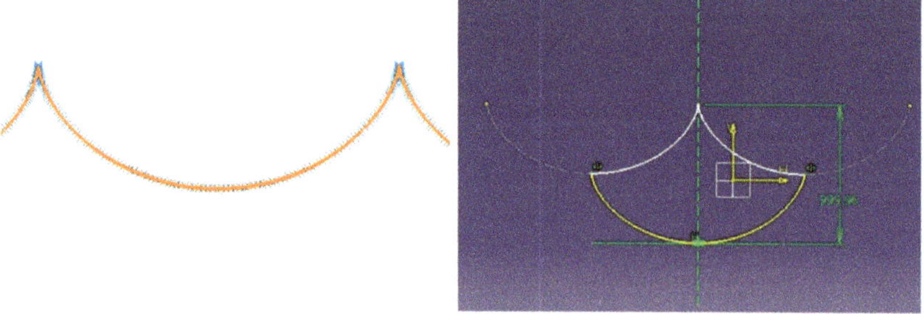

Figure 4. Generation of the cycloid by means of a spline.

Once the cycloidal curve is obtained, and by extrusion of a closed sketch that is created from the cycloid, the 3D models of the two cycloid blades can be obtained (Figure 5).

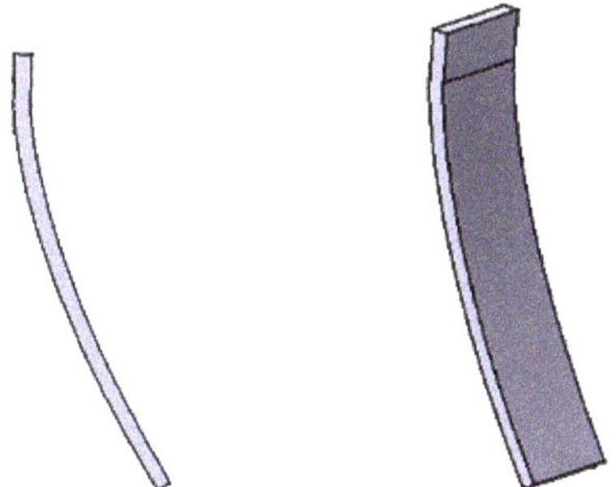

Figure 5. Cycloidal blade.

3.3. Assembly of Subsets and Final Set

This operation was carried out in the CATIA V5 Assembly Design module through the application of a series of geometric and movement restrictions (degrees of freedom) to the different parts, which place them in their final positions. Thus, once each of the subsets was assembled, the final set could be assembled.

Among all the subsets, the pendulum is perhaps the most complicated. Since the pendulum is suspended by two strings between the cycloidal blades (Figure 6), and since CATIA V5 cannot model flexible elements such as the string, this difficulty was resolved by assimilating it into a chain formed by seven links, which represents an acceptable approximation of the real behaviour of the string. Figure 7 shows the pendulum geometry tree, where the various parts of this subset can be observed.

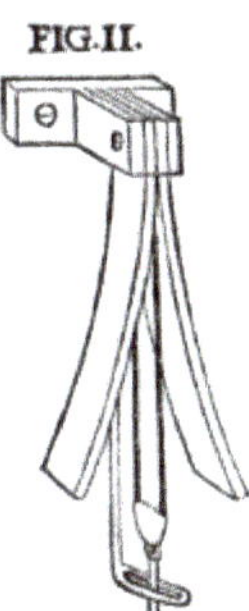

Figure 6. Pendulum between cycloidal blades according to *Horologium Oscillatorium*.

CATIA V5 considers each sub-assembly in a rigid way; if it becomes flexible and becomes an independent mechanism, then the possibility of relating it to another mechanism is lost, which is inconvenient for kinematic simulation.

Likewise, the model of the chain (simulated string) was given a real look (Figure 8), in order to attain a satisfactory result.

Finally, before obtaining the complete assembly of the isochronous pendulum clock (Figure 9), various sub-assemblies were made, such as that of the spheres, the second disk, and the pendulum.

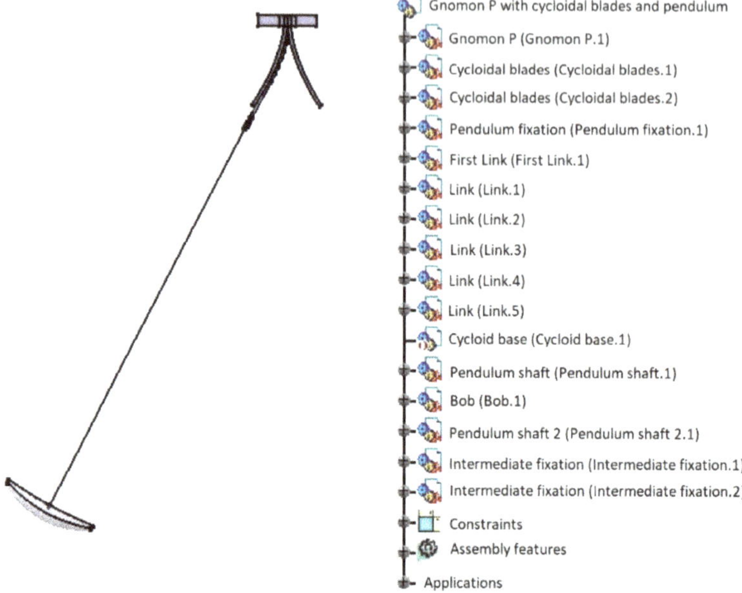

Figure 7. Pendulum geometry tree.

Figure 8. Rendering of the chain with a real string appearance.

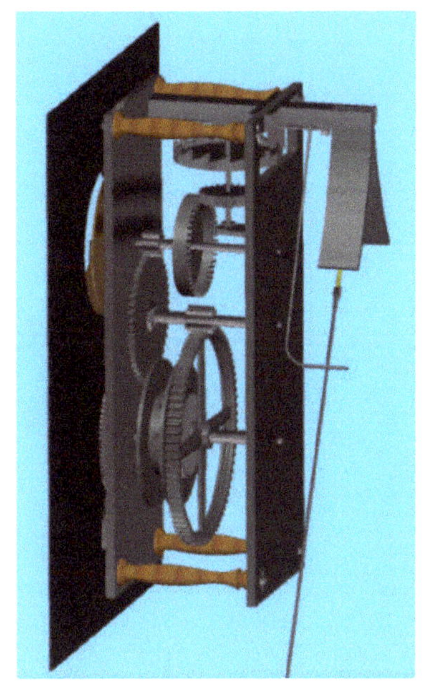

Figure 9. Rendering of the complete assembly of the isochronous pendulum clock.

4. Simulation of Pendulum Kinematics

For the kinematic simulation of the pendulum (the movement of which posed the most problematic phase), the DMU Kinematics Simulator module of CATIA V5 was applied. This module requires the existence of a fixed part for the simulation of a mechanism, and it was established at the beginning as a restriction, and subsequently the unions between the different parts were defined. Without doubt, the simulation of the movement of the pendulum presented the greatest difficulty.

This simulation raised two main drawbacks. In the first place, CATIA V5 fails to allow flexible elements, such as the string, to be modelled, which has been solved by approaching the string as a chain of seven links. Secondly, the possibility that the chain adapts to the cycloidal blades in its periodic movement should also be borne in mind. CATIA has the capability for contact detection, but if this option is activated, then at the moment it detects a contact, the simulation would halt.

The movement of the pendulum was resolved in two ways: By means of commands, that is, by conveniently moving each degree of freedom, and by means of functions, that is, by imposing a function on each degree of freedom of the driving mechanism. Therefore, since the string was modelled as a seven-link chain, at each point of connection between the links there is a degree of freedom of rotation, which are the drivers of the pendulum movement.

By means of command simulation (Figure 10), one of the properties of the cycloid discovered by Huygens is verified: "The evolute of a cycloid is the displaced cycloid itself", since it can be observed that the cycloid of the wooden base and the cycloid of the cycloidal blades are the same but displaced.

Figure 10. Command simulation.

To this end, the simulation of a single movement period is edited with commands. One by one the degrees of freedom are moving, in the most similar way to reality, in the 'Kinematic Simulation' window. At the same time, in the 'Edit Simulation' window, frames are successively inserted for each movement of a degree of freedom made.

Subsequently, the commands were gradually moved, so that when they reach the cycloidal blades, the links adapted to them as much as possible, thereby ensuring, as far as possible, the tangency of the links to the corresponding cycloidal blade, in such a way that the behaviour closely resembled that of a string. The simulation was then compiled, and an animated film generated, thereby obviating the need for the computer to create it every time a visualization is desired, which would greatly slow down the movement.

Subsequently, the generated animation was cyclically reproduced in order to attain the periodic movement of the pendulum, since only one period of the movement was inserted, and the film is captured in a video.

Finally, in this video it can be seen that the property enunciated by Huygens is fulfilled and that the pendulum model is a very good approximation for the result obtained, although a small error is introduced due to the existence of the clamp that must be used for the chain to exit between the cycloid blades, which means that a small part of the cycloid is neglected (Figure 11).

Figure 11. Detail of the chain clamp between the cycloidal blades.

By means of simulation through functions, the animation of the clock as a whole is made possible; however, due to the high number of parts that each move at a different speed, and with respect to the other parts, a discrete synchronization by means of commands of the complete mechanism becomes practically impossible.

When a mechanism is created with CATIA V5, the reference between elements is set as the position they have at the time of creating the joint, and hence, before creating the joints in the Assembly Design module, the links are placed tangentially to one of the cycloidal blades (Figure 12), and it is at that moment when the joints between the different links are created.

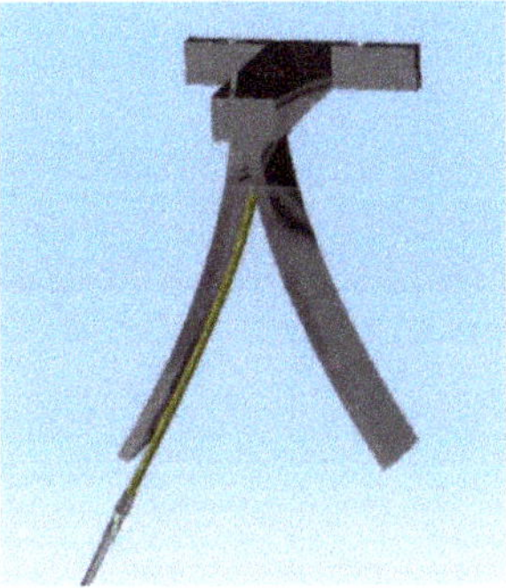

Figure 12. Position of the links tangential to the cycloid blade.

The function (Figure 13) that describes the movement of a simple pendulum (simple periodic movement or simple harmonic movement) is [41]

$$y\,(t) = A \sin\,(\omega t + \Phi)$$

where A is the amplitude of the periodic movement and Φ is the offset.

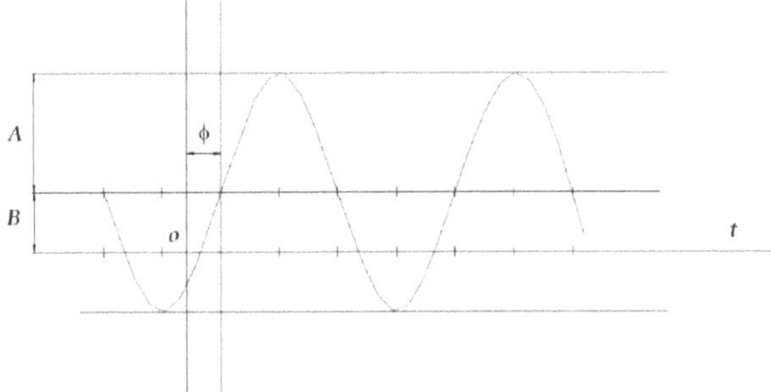

Figure 13. Function describing the movement of the simple pendulum.

For the pendulum to start moving at its point of maximum amplitude ($\Phi = \pm\,90°$), and since the pendulum must mark seconds, ω must be 180°/s.

In this case, the chain of seven links has been modelled as a composite pendulum, in which each link describes a simple periodic movement with respect to the previous link. Therefore, the angular functions that simulate the movement of these links are

$$\text{For the first link: } \theta_1\,(t) = \theta_1 \sin\,(180t - 90) + \theta_1$$

$$\text{For the remaining links: } \theta_j\,(t) = (\theta_j - \theta_i) \sin\,(180t - 90) + (\theta_j - \theta_i)$$

where j is a link and i is the link immediately before. In order to fully define the links, it suffices to measure the angles with respect to the vertical in the initial position of tangency to the cycloidal blade (Figures 14 and 15).

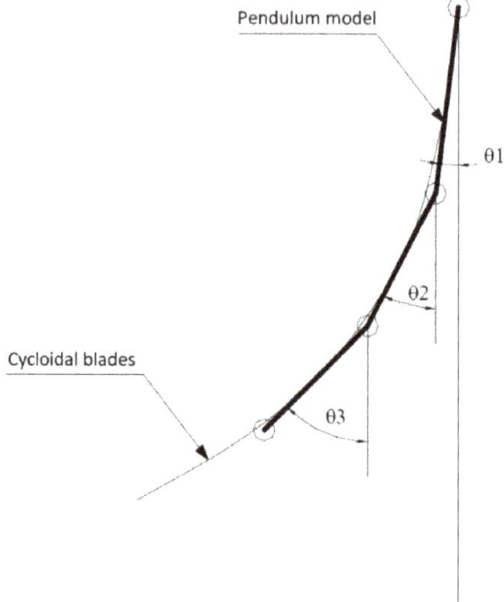

Figure 14. Modelling of the seven-link chain as a composite pendulum.

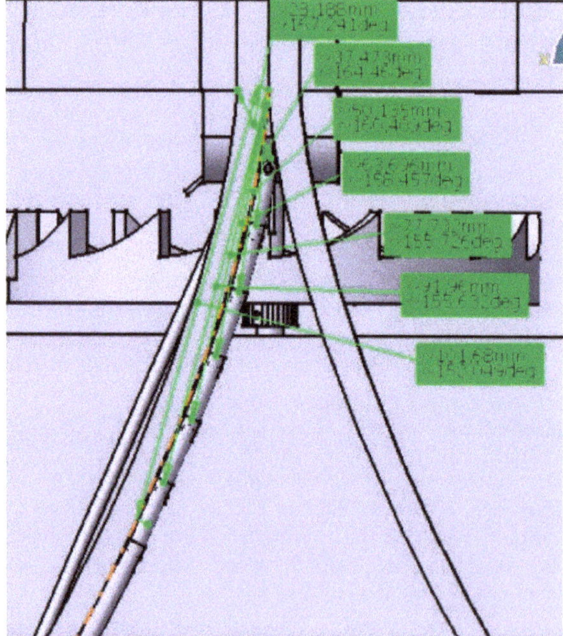

Figure 15. Values of the angles that form the links with respect to the vertical in their initial position of tangency to the cycloidal blade.

Therefore, the laws that describe the movement of the links are:

$$\theta_1(t) = (180 - 167.241)\sin(180t - 90) + \theta_1 = 12.759\sin(180t - 90) + \theta_1$$

$$\theta_2(t) = (167.241 - 164.460)\sin(180t - 90) + \theta_1 = 2.781\sin(180t - 90) + \theta_1$$

$$\theta_3(t) = (164.460 - 160.409)\sin(180t - 90) + \theta_1 = 4.051\sin(180t - 90) + \theta_1$$

$$\theta_4(t) = (160.409 - 158.457)\sin(180t - 90) + \theta_1 = 1.952\sin(180t - 90) + \theta_1$$

$$\theta_5(t) = (158.457 - 155.726)\sin(180t - 90) + \theta_1 = 2.731\sin(180t - 90) + \theta_1$$

$$\theta_6(t) = (155.726 - 155.632)\sin(180t - 90) + \theta_1 = 0.094\sin(180t - 90) + \theta_1$$

$$\theta_7(t) = (155.632 - 152.049)\sin(180t - 90) + \theta_1 = 3.583\sin(180t - 90) + \theta_1$$

Finally, the process that follows is analogous to the previous process: In the simulation with commands a simulation is edited, and a film is then created or a video is generated. As can be observed in Figure 16, the string is curved before reaching the cycloidal blade, and precisely at the moment of arrival it adopts the shape of the blade, which represents a good approximation.

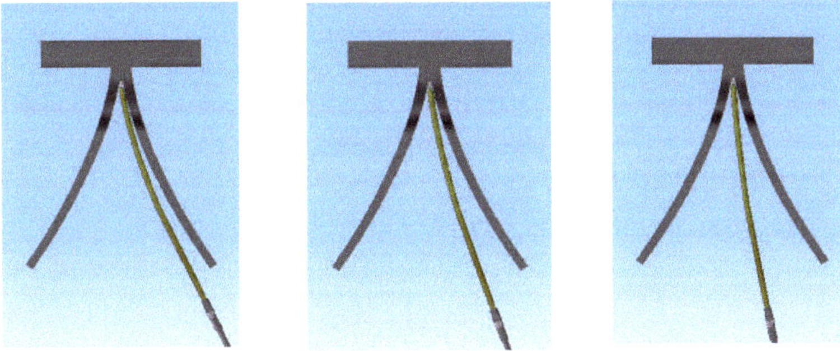

Figure 16. Sequence of chain curvature (simulated string) before reaching the cycloidal blade.

In the screenshots of the animation of Figure 16, the good approximation of this model can be appreciated: The chain (simulated string) is curved before reaching the cycloidal blade, and, exactly at the moment of arrival, its form is adopted. Although the chain does not move exactly like a string, the pendulum shaft carries the necessary movement to regulate the march of the clock. The reason why a greater number of links was not been chosen, which in principle would appear to guarantee a greater reliability of the model, lies in dimensional and geometric causes. From the dimensional point of view, the pendulum section that was represented (which was small in size), could not be easily represented with more than seven links, while from the geometric point of view, seven links sufficiently rectified the cycloid, a curve to which the pendulum had to adapt geometrically. Thus, seven links provides a working scenario of relative comfort and, as seen in the videos and animations obtained, the pendulum describes the cycloid very closely.

Furthermore, the kinematic relationships between the various gears are established by their transmission ratios and it is, therefore, only necessary to establish motion simulation functions for the remaining degrees of freedom. Once these degrees of freedom have been set, the clock can then be simulated and various animations of its operation can be generated.

5. Conclusions

In this article, the three-dimensional modelling of the isochronous pendulum clock of Christiaan Huygens is presented, together with the kinematic simulation of the pendulum using the CATIA V5 software.

The 3D model of the isochronous pendulum clock has been obtained with exactly the same proportions as those described in the work *Horologium Oscilatorium*, thanks to the sizing criteria adopted. However, the only information available regarding the operation of the pendulum clock appears as a brief description of the parts that compose it, the number of teeth of each wheel, the speed at which each axle should move, and the dimensions that the pendulum should have to mark the seconds. There is no information regarding the shape or dimensions of each piece, and hence the design proposed in this research has been created with only the graphic information available in Figure 1. Specifically, given the absence of detailed information regarding the dimensions of the elements, it has been assumed that the gear train as a whole measured vertically one third or, at most, half the length of the pendulum, which has enabled the primitive diameter of the largest wheel (crown wheel), C, of the gear train to be estimated as being 240 mm. From this dimension, it has been possible to define the dimensions of the rest of the wheels and the positioning of the axles, since the dimensioning of a pendulum clock is totally parametric, and depends solely on the length of the pendulum, which univocally implies a relationship of gears.

On the other hand, the main challenges faced in this investigation include the design of the gears of different tooth profiles, the design of the two cycloidal blades, the simulation of the pendulum movement, and, especially, the behaviour of the strings that support the bob of the pendulum. Moreover, one of the main limitations is that presented by CATIA V5 for the modelling of flexible elements such as strings, an inconvenience successfully resolved by approaching the string as a simple chain of seven links that behave as a composite pendulum. The rationale for choosing the number of links lies in dimensional and geometric causes. From the dimensional point of view, the pendulum section that was represented (which was small in size), could not be easily represented with more than seven links, while from the geometric point of view, seven links sufficiently rectified the cycloid, a curve to which the pendulum had to adapt geometrically.

The kinematic simulation of the pendulum was also performed, whereby the exactitude of the clock in the measurement of time was verified and the pendulum was described with close approximation to the cycloid. Furthermore, thanks to CAD and kinematic simulation techniques, it has been possible to understand and interpret the operation of Huygens's isochronous pendulum clock, which in turn has enabled virtual recreations to be generated that can serve as educational tools in the museum of Science History, and has facilitated the dissemination of technical historical heritage. Moreover, future uses may involve the development of applications of virtual reality and augmented reality, the incorporation of the WebGL model into a website, and the ability to print in 3D using additive manufacturing techniques.

Finally, the results of this research regarding the CAD 3D model can be employed by other researchers, and shared online via maker communities, such as GrabCAD [42] and Thingiverse [43].

Author Contributions: Investigation, J.I.R.-S. and F.J.G.-C.; Methodology, F.J.G.-C. and G.D.R.-C.; Validation, F.J.G.-C.; Writing—original draft, J.I.R.-S. and F.J.G.-C.; Writing—review & editing, J.I.R.-S., F.J.G.-C. and G.D.R.-C. All authors have read and agreed to the published version of the manuscript.

Funding: This research received no external funding.

Acknowledgments: We sincerely appreciate the work of the reviewers of this manuscript.

References

1. Baillie, G.H. Huygens' pendulum clock. *Nature* **1941**, *148*, 412. [CrossRef]
2. Lepschy, A.M.; Mian, G.A.; Viaro, U. Feedback-control in ancient water and mechanical clocks. *IEEE Trans. Educ.* **1992**, *35*, 3–10. [CrossRef]
3. Senator, M. Synchronization of two coupled escapement-driven pendulum clocks. *J. Sound Vibr.* **2006**, *291*, 566–603. [CrossRef]
4. Rojas-Sola, J.I.; De la Morena-De la Fuente, E. The Hay inclined plane in Cooalbrookdale (Shropshire, England): Geometric modeling and virtual reconstruction. *Symmetry* **2019**, *11*, 589. [CrossRef]
5. Rojas-Sola, J.I.; Galán-Moral, B.; De la Morena-De la Fuente, E. Agustín de Betancourt's double-acting steam engine: Geometric modeling and virtual reconstruction. *Symmetry* **2018**, *10*, 351. [CrossRef]
6. Rojas-Sola, J.I.; De la Morena-de la Fuente, E. Digital 3D reconstruction of Agustín de Betancourt's historical heritage: The dredging machine of the port of Krondstadt. *Virtual Archaeol. Rev.* **2018**, *9*, 44–56. [CrossRef]
7. Rojas-Sola, J.I.; De la Morena-de la Fuente, E. Geometric modeling of the machine for cutting cane and other aquatic plants in navigable waterways by Agustín de Betancourt y Molina. *Technologies* **2018**, *6*, 23. [CrossRef]
8. Rojas-Sola, J.I.; De la Morena-de la Fuente, E. Agustín de Betancourt's wind machine for draining marshy ground: Approach to its geometric modeling with Autodesk Inventor Professional. *Technologies* **2017**, *5*, 2. [CrossRef]
9. Río-Cidoncha, G.; Martínez-Palacios, J.; González-Conde, L. Torriani's mechanical device for cartying water to Toledo. *Ing. Hidráulica Mex.* **2008**, *23*, 33–44.
10. Pennestri, E.; Pezzuti, E.; Valentini, P.P.; Vita, L. Computer-aided virtual reconstruction of Italian ancient clocks. *Comput. Animat. Virtual Worlds* **2006**, *17*, 565–572. [CrossRef]
11. Herrmann, J. *Atlas de Astronomía*; Alianza: Madrid, Spain, 1987. (In Spanish)
12. Moore, P. *Astronomía*; Vergara: Barcelona, Spain, 1963. (In Spanish)
13. Clagett, M. *Ancient Egyptian Science: Calendars, Clocks, and Astronomy*; American Philosophical Society: Philadelphia, PA, USA, 1995; Volume 2.
14. Arago, F. *Grandes Astrónomos Anteriores a Newton*; Espasa-Calpe: Buenos Aires, Argentina, 1962. (In Spanish)
15. Model of Huygens' Pendulum Clock, as Illustrated in His 'Horologium' of 1658. Available online: https://collection.sciencemuseumgroup.org.uk/objects/co817/model-of-huygens-pendulum-clock-as-illustrated-in-his-horologium-of-1658-pendulum-clock-reconstruction (accessed on 24 December 2019).
16. Tait, H. *Clocks and Watches*; Harvard University Press: London, UK, 1983.
17. Wenham, E. *Old Clocks*; Spring Books: London, UK, 1951.
18. Britten, F.J. *Watch & Clockmakers' Handbook. Dictionary and Guide*; Antique Collectors' Club Ltd.: Suffolk, UK, 1993.
19. Bennett, M.; Schatz, M.F.; Rockwood, H.; Wiesenfeld, K. Huygens's clocks. *Proc. R. Soc. A* **2002**, *458*, 563–579. [CrossRef]
20. Klarreich, E. Huygens's clocks revisited. *Am. Sci.* **2002**, *90*, 322–323.
21. Huygens, C. *Horologium. 1658*; An English Translation Together with the Original Latin Text in Facsimile by Ernest, L. Edwardes; Antiquarian Horological Society: Ticehurst, UK, 1970.
22. Edwardes, E.L. *The Story of the Pendulum Clock*; Sherratt and Son: Altrincham, Manchester, UK, 1977.
23. Huygens, C. *Horologium Oscillatorium, siue, De motu Pendulorum ad Horologia Aptato Demonstrationes Geometricae*; Facsimile Reprint of the Text in Latin of Paris: Apud, F. Muguet, 1673; Culture et civilisation: Bruxelles, Belgium, 1966.
24. Huygens, C. *Horologium Oscillatorium*; An English Facsimile Reprint of the 1673 Paris Edition; Dawsons: London, UK, 1966.
25. Huygens, C. *Christiaan Huygens' the Pendulum Clock or Geometrical Demonstrations Concerning the Motion of Pendula as Applied to Clocks*; Translated into English of Horologium Oscillatorium with Notes by R.J. Blackwell and Introduction by H.J. Bos; The Iowa State University Press: Ames, IA, USA, 1986.
26. Popular Mechanics: Leonardo da Vinci's Mechanical Lion. Available online: https://www.popularmechanics.com/technology/a29020685/leonardo-da-vinci-mechanical-lion-display (accessed on 24 December 2019).
27. Betancourt's Project. Available online: http://fundacionorotava.es/betancourt/machines (accessed on 24 December 2019).

28. Juanelo Turriano's Device to Raise the Water from the Tagus River to the City of Toledo (Spain). Available online: https://www.youtube.com/watch?v=_qTYt6drnOU (accessed on 24 December 2019).
29. Antelo's Clock: The Clock of the Cathedral of Santiago (Spain). Available online: https://parpatrimonioytecnologia.wordpress.com/2016/05/23/el-reloj-de-antelo-santiago-de-compostela-modelado-texturizado-y-visor-3d-galicia100/#more-2409 (accessed on 24 December 2019).
30. Virtual Reconstruction of the Visigoth Church of Santa María de Melque (Toledo, Spain). Available online: https://www.artstation.com/artwork/v1DD0d (accessed on 24 December 2019).
31. Virtual Reconstruction of the Church of San Agustín de la Laguna (Tenerife, Spain). Available online: https://www.youtube.com/watch?v=QEM_ZV3vLPU&feature=emb_title (accessed on 24 December 2019).
32. Cozzens, R. *Advances CATIA V5 Workbook: [Releases 8 & 9]*; Schroff Development Corporation: Mission, KS, USA, 2002.
33. Saorin, J.L.; Lopez-Chao, V.; De la Torre-Cantero, J.; Diaz-Aleman, M.D. Computer-aided design to produce high-detail models through low cost digital fabrication for the conservation of aerospace heritage. *Appl. Sci.* **2019**, *9*, 2338. [CrossRef]
34. Rojas-Sola, J.I.; De la Morena-de la Fuente, E. The Hay inclined plane in Coalbrookdale (Shropshire, England): Analysis through computer-aided engineering. *Appl. Sci.* **2019**, *9*, 3385. [CrossRef]
35. Rojas-Sola, J.I.; De la Morena-de la Fuente, E. Agustin de Betancourt's double-acting steam engine: Analysis through computer-aided engineering. *Appl. Sci.* **2018**, *8*, 2309. [CrossRef]
36. Tashi; Ullah, A.M.M.S.; Watanabe, M.; Kubo, A. Analytical point-cloud based geometric modeling for additive manufacturing and its application to cultural heritage preservation. *Appl. Sci.* **2018**, *8*, 656. [CrossRef]
37. Pezzano, P.; Klein, A. *Serie Elementos de Máquina: Engranajes y Poleas*; Librería el Ateneo: Buenos Aires, Argentina, 1980; Volume III. (In Spanish)
38. Carrera, T. *Engranajes: Trazado Teórico y Práctico. 1ª y 2ª Parte*; Carrera Soto: Sevilla, Spain, 1987. (In Spanish)
39. Merrit, H.E. *Gear Engineering*; John Wiley & Sons: New York, NY, USA, 1971.
40. Litvin, F.L. *Gear Geometry and Applied Theory*; Prentice Hall: Englewood Cliffs, NJ, USA, 1994.
41. Maiztegui, A.P.; Sábato, J.A. *Introducción a la Física*; Kapeluz: Buenos Aires, Argentina, 1966. (In Spanish)
42. GrabCAD: Free CAD Library. Available online: https://grabcad.com/library (accessed on 24 December 2019).
43. Thingiverse: Models. Available online: https://www.thingiverse.com/explore/newest/models (accessed on 24 December 2019).

 *applied sciences*

Article

The Contribution of the Segovia Mint Factory to the History of Manufacturing as an Example of Mass Production in the 16th Century

Francisco García-Ahumada * and Cristina Gonzalez-Gaya *

Department of Construction and Manufacturing Engineering, ETSII-Universidad Nacional de Educación a Distancia (UNED), C/Juan del Rosal 12, 28040 Madrid, Spain
* Correspondence: fgarcia1895@alumno.uned.es (F.G.-A.); cggaya@ind.uned.es (C.G.-G.); Tel.: +34-629-67-27-37 (C.G.-G.)

Received: 1 October 2019; Accepted: 26 November 2019; Published: 7 December 2019

Abstract: A new means of minting currency was first used at the Hall Mint in Tyrol in 1567. This new minting process employed a roller instead of a hammer and used hydropower to fuel the laminating and coining mills, as well as ancillary equipment, such as the forge or the lathe. In 1577, Philip II of Spain expressed his interest in the new technology and, after a successful technology transfer negotiation with the County of Tyrol, Juan de Herrera was commissioned to design a factory to accommodate this new minting process. The resulting design seamlessly integrated this new technology. The architectural layout of the factory was derived from the integration of different trades related to the manufacturing workflow, and their effective distribution within a more effective workplace allowed for better use of the hydraulic resources available, and, thus, improvements in the productivity and reliability of the manufacturing process, as well as in the quality of the finished product. Juan de Herrera's design led to the creation of a ground-breaking manufacturing process, unparalleled in the mint industry in Europe at the time. Segovia Royal Mint Factory (SRMF), as one of the first examples of mass production in the proto-industrial stage, represents a historic landmark in its own right. The objective of this article is to analyse the design of the SRMF to highlight its main innovations. For this purpose, the abundant literature on this project will be reviewed.

Keywords: mint industry; construction; transfer technology; align process; architectural layout

---

1. Introduction and Goals

The Segovia Royal Mint Factory (SRMF) has its origins in King Philip II of Spain's demand for a founder for the manufacture of cannons. In a letter dated 23 October 1574, the King stated that "«In these Realms there is a need for a couple of good artillery founders that they tell me tend to live in Nuremberg … »" [1]. In 1577, this search led the King to the discovery of a new minting process, which had replaced the former minting process that involved using a hammer.

This new minting process was developed at Hall Mint, in 1571, and used a roller instead of a hammer. The main advantage of this new manufacturing process, and what made it so attractive to the crown of Spain, was its ability to maintain uniform weight, thickness, quality of type, engraving and precision parameters in every coin produced, which, in turn, made illicit activities such as clipping and filling incredibly difficult.

Figure 1 presents two images. The first corresponds to a coin minted (Real de a ocho) in the mechanized SRMF of Segovia in 1627, and the second to a "Real de a ocho", minted by a hammer, in the Casas Vieja (The effects of the clipping on the coin can be seen, in relation to the left coin) mint of Segovia, in 1627 [1–3].

Figure 1. Comparison between two coins minted in the Segovia Royal Mint Factory (SRMF) and minted by a hammer (Based on [1–3]).

Negotiations between the County of Tyrol and the Kingdom of Spain, regarding the transfer of technology, were initiated in 1581. The final approval of the agreement, allowing the introduction of the roller-based minting process in Spain, is detailed in a letter located in Innsbruck, dated 4 February 1582. In it, Archduke Ferdinand, cousin of King Philip II of Spain, gives Baron Khevenhüllern, Imperial Ambassador at the court of Philip II, permission to communicate the success of the negotiations to the Crown of Spain [2].

The SRMF project began development in 1582, during what is known as the proto-industrialization stage [4]. The SRMF would become one of the first European factories. It stayed operational from 1586 to 1866. In 2012, it became a museum.

Several innovations appear during the development of this project, which make it unique in the history of manufacturing. The most relevant are:

i.   *Technology transfer process.* The SRMF Project is based on a technology transfer agreement between two nations.

ii.   *Multinational project teams.* In this project, they had to coordinate a multinational team for its development and operation.

iii.   *Transportation Logistics.* This project required a logistic transport model for the supply of machines and engines for manufacturing.

iv.   *Architectural layout.* An architectural design aligned with the mass manufacturing process, two centuries before the industrial revolution at the time of proto-industrialization [4,5].

v.   *Hydraulic model.* A hydraulic model to supply energy for the manufacturing process.

In the following sections, the following aspects of the SRMF Project will be analysed:

- The scope of the work and the project team.
- The stages of the project.
- The design and construction of the SRMF, serving as an example of an alignment between manufacturing methods and architectural layout.

The focus of the article is to analyse the different innovations of the SRMF project.

2. Materials and Methods

The material used is the existing bibliography of the sections mentioned above, which has been selected based on the analysis of the innovations already mentioned.

The methodology is formed of an analysis of the verified facts. In the analysis of the bibliography of this project, we have searched for the relevant innovations throughout the history of the industry.

3. Results

The order of the results will follow the order indicated in the introduction.

3.1. The Scope of Work and the Project Team

The Segovia Royal Mint Factory project started with an international technology transfer agreement, involving the County of Tyrol—under the control of Archduke Ferdinand—and the Kingdom of Spain—under the rule of King Philip II. The scope of the SRMF project can be divided into two parts.

The first part is formed of the transfer of the roller-based minting process, the manufacture of the rolling mills and other necessary machinery at the Hall Mint in Tyrol and the transportation of the equipment, along with its operators, to Spain.

The technology transfer agreement is notable, since it is a contract between two countries in which a technology is transferred through its equipment and tools, as well as the workers necessary to its operation, as skilled Austrian workers accompanied the machinery on its trip to the Kingdom of Spain.

The second part is formed of the design and construction of the SRMF facilities, as well as the selection of a suitable site for the factory. The factory had to be comprised of a building, or a series of buildings, which were able to accommodate all the elements necessary for autonomous operation, including all the mechanical equipment involved in the manufacturing process powered by hydraulic energy.

After the signing of the technology transfer agreement in 1582, a team of experts from both countries were gathered to develop and manage the SRMF project. The Crown of Spain provided:

- Juan de Herrera, appointed as project leader due to his experience in both architectural and engineering projects. Of particular relevance to this project was his previous experience in water conveyance and hydraulic networks, as well as his participation, along with Jacome da Trezzo, in 1579, in the design and construction of the Jasper Mill, which was used to cut hard stones for the Monastery of El Escorial [6].
- Francisco Ribera, the "Veedor", or Crown's Representative, in the project, and in charge of the project's economic aspects.

The County of Tyrol provided a team of six German technicians: a master teacher, three carpenters, a blacksmith and a locksmith. They were sent to Spain to provide technical support from the early stages of the project, in fulfilment of the terms of the technology transfer agreement [7].

The launch of the project began with the arrival of the German technicians in Spain, in April 1582.

3.2. The Stages of the SRMF Project

In the following sections, the different stages of the SRMF project will be analysed.

3.2.1. Site Selection

Although the different criteria for site selection are not analysed in this paper, it is worth pointing out the importance of the availability of a constant flow of water throughout the year. Consequently, Segovia was the chosen location, despite Seville being the entry point for the precious metals coming from the Americas, which seemed a more logical option at first. More information on the subject can be found in [8].

The selected plot was located adjacent to the Santa Maria del Parral Monastery, along the Eresma river. A mill used to produce paper was already situated in this location—the Mill of San Millan. According to the Land Registry (reference number 5145601VL0354N), it had a total area of 7635 m^2.

3.2.2. Manufacture of the SRMF's Machinery

The transportation of machinery and ancillary equipment from the Hall to Segovia was one of the most challenging stages of the SRMF project, when considered from a logistical perspective [9]. Given the distance (approximately 2000 km) between both cities, the difficult orography of some of the areas, and the large amount of equipment needing to be transported—25 carts were necessary, once everything was carefully packaged, to avoid any damage—the Spanish Road [10] was the chosen route, for safety reasons. Some members of the Hall Mint staff—an assayer, an engraver, a coin master, a founder and four coin technicians—were sent along for equipment installation, operation and personnel training purposes.

The journey was divided into three legs. Starting in Hall, the first leg was overland and had three stopovers: Como, Milan and Genoa. The second leg was the journey between Genoa and Barcelona, and was done by sea, while the last part of the trip, from Barcelona to Segovia, was an overland journey. The journey took from October 1584 to June 1585 [9]. Due to its magnitude, this trip can be regarded as a milestone in the history of logistics.

3.2.3. Start-Up Tests and First Minting

With the arrival of the equipment in June 1585, and the team of technicians that travelled with it, the installation of the machinery in the SRMF facility began, quickly followed by start-up tests and the first minting (shown in Table 1).

Table 1. Summary of the start-up tests and first minting (from [9], p. 101).

Test Summary		
Date	**Classification of Glenn [1]**	**Documentary classification**
jul-85	First Test	First test (cooper)
dec-85	Second and third test	First test (lost silver) and second test (36 coins)
Mar-86	First work	First test (1489 marks: 100 coins to the King)

Once the operation procedure was implemented and refined, the members of Hall Mint proceeded to transfer their knowledge to local personnel. Their technology was replicated in all the minting factories and abandoned in the period 1660–1664. Only the rolling process persisted through the transition of the coining press into flywheel presses [9].

3.3. The Design and Construction of SRMF

The SRMF was among the first mechanized factories dedicated to coin minting in Europe in the 16th century, and the first of its kind in the Kingdom of Spain. It was built on a plot located by the Eresma River in Segovia. The plot had an existing building, an old mill used to produce paper, and a total surface area of 7635 m^2. The requirements of the project consisted of a facility able to accommodate all the elements necessary to operate autonomously, with all the mechanical equipment involved in the manufacturing process powered by hydraulic energy.

Juan de Herrera employed an innovative but rather unusual approach in the design of the architectural layout of the SRMF; he analysed the new minting process in depth and, relying heavily on the technical advice provided by the support team of German technicians, integrated the different trades related to the manufacturing workflow and arranged them in an effective workplace distribution, which allowed for better use of the hydraulic resources available and generated significant reductions in both cost and time.

In the following sections, we will review the minting process by roller and how it relates to the architectural layout designed by Juan de Herrera, as well as the hydraulic system designed to power the mechanical elements of the factory.

3.3.1. The Minting Process by Roller and the SRMF Architectural Layout

The different activities related to the minting process by roller can be grouped into the following categories:

- Intrinsic: All activities related to the process of creating alloys and casting the alloy strips from which coins were manufactured.
- Extrinsic: Tasks associated with the mechanical, thermic and chemical treatments applied to the alloy strips used in the manufacturing of currency.
- Auxiliary: Includes activities connected to machinery maintenance, as well as engraving, forging and lathing processes.
- Administrative: This category can be divided into two subgroups. The first includes all activities related to the guard and custody of raw materials (metal ingots), alloy strips, coins and metal alloy scraps generated by the blank cutting process, while the second encompasses all activities linked to the registration and verification of the weight and quality (also called Law) of the metals employed to create the alloys, as well as the alloy strips, coins and metal alloy scraps resulting from the blank cutting process.

A list of the roles necessary for the operation of the SRMF, along with a brief description of their corresponding duties, is displayed in Table 2, below.

Table 2. Roles and responsibilities (prepared by the authors, based on [11]).

Position in the Organization	Description
The Metal Owner	Person who delivers the metal for the elaboration of coins
The assayer	Person responsible for monitoring the grade in the incoming metal and the strips
The scribe	Person who was responsible for giving written testimony of all the activities that were carried out in the Royal Segovia Mint
The founder	Person responsible for starting from the metal that were delivered to the Royal Segovia Mint, obtaining by casting the strips
The engraver	Person responsible for designing the roller-die to make coins by rolling
The blacksmith	Person responsible for all the activities of the smithy, forge etc.
Lathe operator	Person responsible for lathe activities
Chief coin master	Person responsible for the extrinsic process
Waterwheel master	Person responsible for the availability of the waterwheels that provided the driving force by water
Treasure	It is until 1730 the maximum responsible for the Royal Segovia Mint

The sequence of activities that make up a cycle in the production of currency at the SRMF, from the beginning—when the raw material, metal, enters the production chain—until the end of the manufacturing process—when the final product, in the form of coins, is obtained—is shown in the flow charts of Figures 2 and 3. Both charts have been based on those shown in [2], on pages 60 and 61.

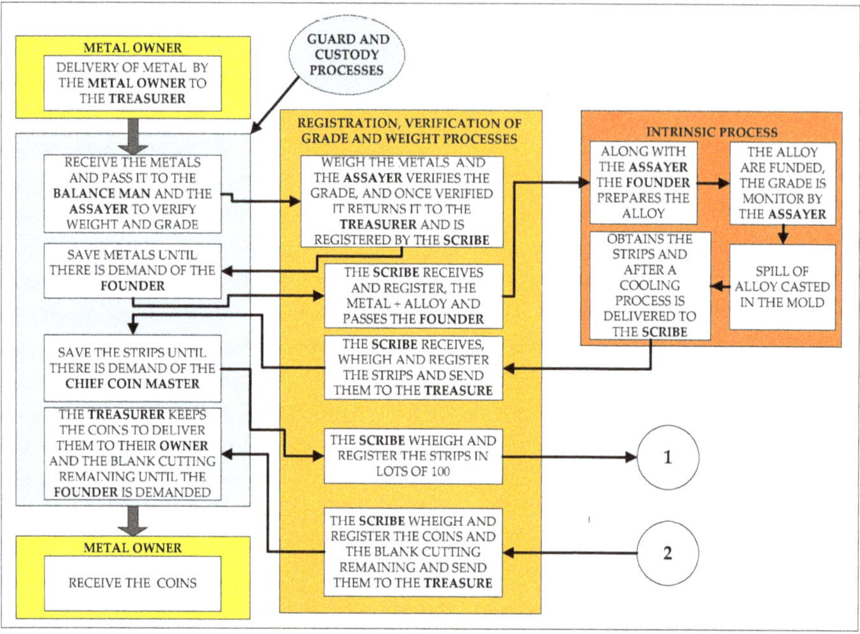

Figure 2. Intrinsic, auxiliary and administrative process (based on [2] pages 60 and 61).

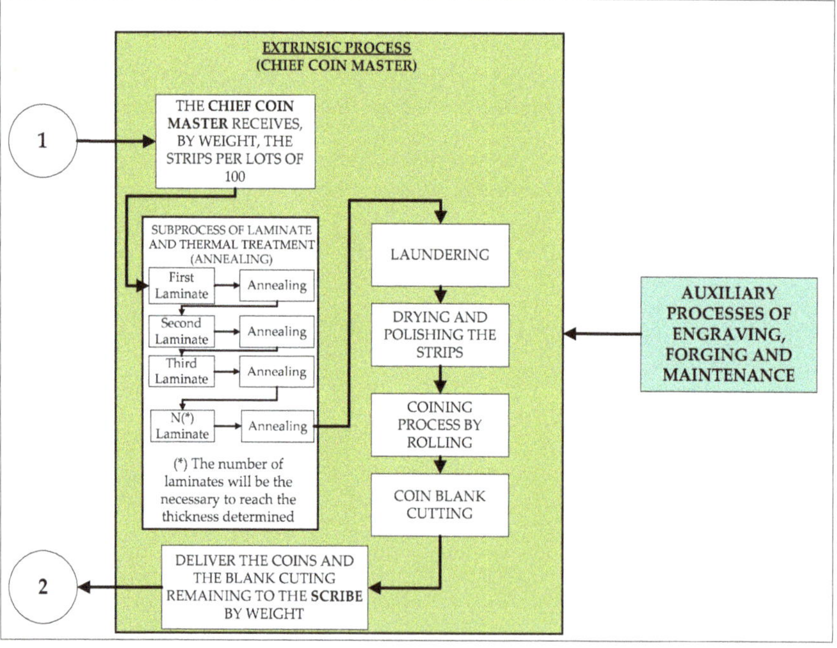

Figure 3. Extrinsic and auxiliary process (based on [2] pages 60 and 61).

The resulting design by Juan de Herrera had a built surface of 4.423 m^2 and was made up of three buildings (see Figure 4). The first was called "Edificio del Patio Alto", and the other two were called

"Ingenio Grande" and "Ingenio Chico". A water system located between the "Ingenio Grande" and "Ingenio Chico" buildings channelled water to power the waterwheels, located along the façade of the "Ingenio Grande".

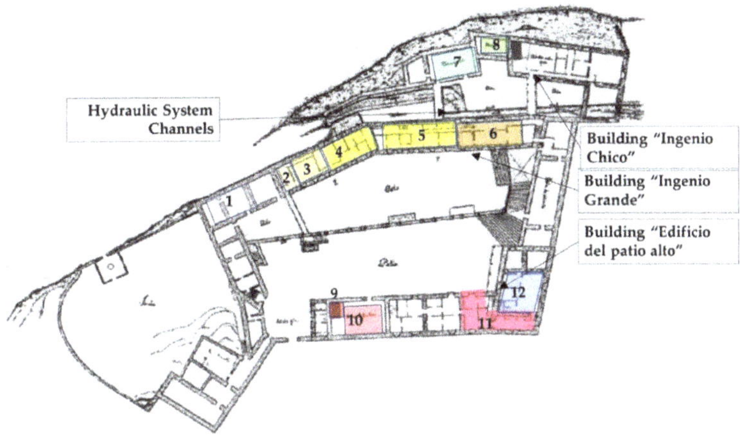

1. Forging (support in the construction phase). Workshop (in operation)
2. Engraving office
3. Blank cutting
4. Coin Rolling mil by roller dies
5. Rolling mill
6. Forge
8. When the strips are blanched and rinsed
9. Heat treatment (Annealing)
10. Scraps
11. Smelting
12. Assay
13. Wheigt Room

Figure 4. SRMF Architectural layout with distribution of uses (Based on [11]).

The activities that made up the currency-minting production chain were organized among the different buildings of the SRMF, as follows:

- Edificio del Patio Alto: This building was the SRMF management headquarters, and, until 1730, it was run by the Treasurer. This was where the input of raw material, (metal ingots) and the output of finished product (coins), took place, and where administrative and intrinsic activities were carried out. From the moment the metal ingots entered the production cycle, each stage of the process was carefully registered and controlled: the registration and verification of the grade and weight of metal used to cast the alloy strips, the alloy strips themselves and the coins obtained from them. From this building, the alloy strips were sent to the "Ingenio Grande" and came back in the form of coins and alloy scraps.
- Ingenio grande: This was the biggest of the three buildings, and it was managed by the Coin Master. Mechanical and auxiliary processes were developed here (see Figure 5).

Figure 5. "Ingenio Grande" (image property of the authors).

The fact that the equipment necessary to perform such activities had to be powered by hydraulic energy affected the arrangement of the different activities within the space. A sequential order was followed, based on the manufacturing process. A series of waterwheels were installed outside—along one of the longer walls—to power the forge blower, the hammer, the lathe, the rolling mill and the coin die machine (see Figure 6).

Figure 6. Functional areas (image property of the authors).

The different functional areas within the building were the following:

- The Forging Shop Area: This stretched along the first three waterwheels, which powered the forge, hammer and lather. It was where the auxiliary processes supporting the manufacturing activity were performed.
- Laminating Rolling Mill Area: This accommodated the rolling mills that were manufactured in the Hall. It was where alloy strips were subjected to successive processes of lamination, in order

to reach the desired thickness. This area was linked to the "Ingenio Chico" building, since the strips were constantly moving between both buildings to alternate the lamination processes with the thermic treatment meant to restore their mechanical properties.

- Coin Rolling Mill Area: This contained a rolling mill with a coin die installed, which was used to print both sides of the coin on the strip simultaneously (see Figure 7).
- A virtual reconstruction of the laminating and coin mill area can be seen in Figure 8.
- Blank Cut Area: This is where the blank cutting press was used to cut the coins out. Once this process was finished, the coins that were obtained and the alloy scraps generated were sent back to "Edificio del Patio Alto".
- Engraving Area: This is where the engraver produced the different stamp punches.
- Mechanics Workshop Area: This area was dedicated to the performance of all maintenance tasks, especially those related to the water system. It was run by the Waterwheel Master.
- Ingenio Chico: This corresponded to the old San Millan Mill that was part of the original site. The building was repurposed to accommodate the thermal and chemical treatments given to the alloy strips, the former being the annealing process and the latter, the laundering process. The annealing was a thermic treatment, aiming to restore the mechanical properties of the alloy strips at the end of each lamination process they were subjected to, while the laundering was a chemical treatment, meant to whiten the alloy strips, which darkened after rounds of lamination and annealing. This building was also run by the Coin Master.

Figure 7. The rolling mill (image property of the authors).

Figure 8. Illustration by Reiner, Valencia ([12] Figure 3, on p. 97).

3.3.2. The SRMF Hydraulic System

The water system's ability to harness the Eresma River flow, in order to convey the hydraulic energy required to power the equipment related to mechanical and auxiliary processes, was key to the successful operation of the SRMF. The goal of Juan de Herrera's water system design was to adapt the existing structure, which had served the old San Millan Mill, to the more demanding requirements of the SRMF.

The original gravity dam, with an arched floor and lateral drainage located in the left abutment, manually regulated by two gates, was modified to incorporate an outlet meant to divert enough of the Eresma River flow to actuate the waterwheels placed along the façade of the Ingenio Grande (see Figures 9 and 10). The surplus capacity of the bypass channel was designed to be poured by coronation (see Figures 11 and 12). This system was intended to reduce the number of waterwheels used, in case of a decrease in flow availability caused by intense drought [13].

Figure 9. Forging shop waterwheels and the channels. (Image property of the authors).

Figure 10. Forge blower waterwheel. (Image property of the authors).

Figure 11. (Image property of the authors).

Figure 12. Entrance of the water flow in the SRMF. The right opening connects to the channel that feeds the waterwheels, while the left one returns the surplus water to the Eresma River. (Image property of the authors).

The Waterwheels

The wheels were designed to supply the power necessary to move the manufacturing and forging chop equipment of the Ingenio Grande:

"All the wheels were made of timber with the joints and fittings fixed in place by means of wooden wedges, reinforcements and iron nails" [12]. The wheels were back-shot waterwheels. The type of driving waterwheels used had blade-shaped blades. According to *Los veintiún libros de los ingenios y las máquinas* [14], such technology was known in Spain in the period.

The waterwheel operation was as follows (see Figure 13): The water from the dam, flowing from the channel and through the slopping duct (regulated by a gate), reached the wheel blades and communicated an angular speed to the wheel. Once the water had passed through the wheel, it passed to the lower channel, which returned it to the river. It is worth pointing out how remarkably the design maximized the kinetic energy of the fluid reaching the waterwheels (based on height difference (h). The efficiency did not exceed 35% [11].

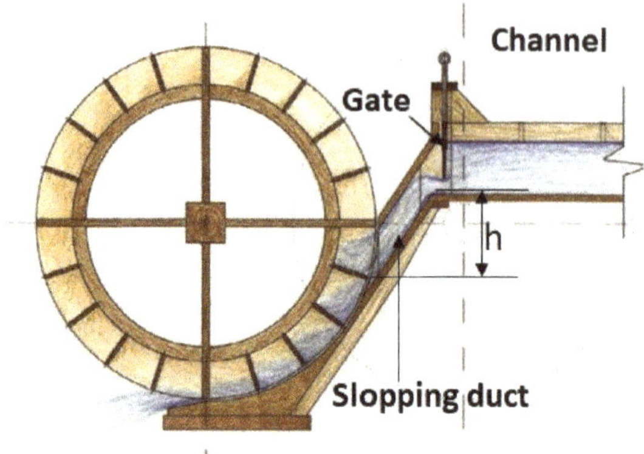

Figure 13. Cross section of the waterwheel for the laminating rolling mill. (Illustration by Reiner, Valencia [12] p. 112).

The powers and size of the waterwheels for the different equipment have been estimated in [11] and can be seen in Table 3.

Table 3. Prepared by the authors, based on [11].

WATERWHEELS								
Functional Area in "Ingenio Grande"	Engine of.	Efficiency Estimate (%) (*)	Power (kW)	Waterwheel Diameter (m)	Number of Blades	Water Flow (l/s)	rpm	Notes
Forging Shop	Forge blower	14%	1.54	2.5	16	62	25	
Forging Shop	Lathe	14%	1.54	2.5	16	62	25	
Forging Shop	Hammer	14%	3.72	2.5	16	146	25	For 100 blows per minute
Laminating rolling mill	Laminating rolling	15.75%	2.05	3.76	20	83	14	
Coin Rolling mill	Coin rolling	15.75%	1.03	3.76	20	42	14	

(*) The efficiency is the waterwheel and the equipment (Hammer, Lathe, etc.).

To model the operation of the waterwheels by CAD (Computer Aided Design), the methodology developed in [15] could be used.

4. Conclusions

In relation to the SRMF Project, these conclusions will focus on the innovative aspects of this project, which are divided into those innovations in the development of the manufacturing model and in the development of the hydraulic model, applicable to the extrinsic and auxiliary processes.

In relation to the development of the new manufacturing model, the following are proposed:

i. In relation to *project management*: What started as an international technology transfer agreement between the Kingdom of Spain and the County of Tyrol became an extraordinary joint effort, where a multidisciplinary team of experts from both countries managed to complete the project in a relatively short period of time, accomplishing a feat of logistics that was unprecedented at that time. It should be mentioned that to ensure the success of the project, part of the Austrian team that participated in the first stage of the life cycle remained in the Kingdom of Spain to perform the subsequent operation of the SRMF [9].

ii.   In relation to the ***architectural layout*** of the SRMF: This was the greatest innovation of this project, which occurred two centuries before the industrial revolution. This innovation must be noted with consideration of the following:

    a.   For the first time, the production model was not installed in existing buildings, but rather, some buildings were designed so that all production processes were accommodated inside.

    b.   The design of the buildings corresponded to the different manufacturing processes; the intrinsic, auxiliary and administrative processes took place in separate, fit-for-purpose buildings. In the buildings where extrinsic and auxiliary processes took place, the intelligence of the design meant that the buildings were operational from 1586 until the factory closed in 1868, with small changes.

    c.   Overall, this demonstrates a perfect integration between technology, distribution of different workplaces and optimal use of hydraulic energy.

iii.   In relation to the *hydraulic model*, the SRMF optimally used hydraulic energy in both the extrinsic and axillary processes of forging and maintenance.

The consequences of these innovations in the production process were mainly the following:

In relation to the product—the coins—a substantial improvement in the quality and homogeneity of the finished product was achieved, and this high standard of the finished product allowed fighting illicit activities, such as clipping and filling, effectively.

In relation to the improvement of productivity, already by 1588, Linggahöl had compared the factories of Segovia and Seville, and considering the blank cutting process and a daily consumption of 250 kg of silver, established that in the SRMF this process was carried out by eight people, estimating that 100 people would be required in Seville [16].

The SRMF is one of the first examples of mass production in the age of proto-industrialization.

Author Contributions: Conceptualization, F.G.-A. and C.G.-G. Methodology, F.G.-A. and C.G.-G. Validation, F.G.-A. and C.G.-G. Formal analysis, F.G.-A. Investigation, F.G.-A. Resources, C.G.-G.; Data curation, F.G.-A. Writing—original draft preparation, F.G.-A. and C.G.-G. Writing—review and editing, F.G.-A. and C.G.-G. Supervision, C.G.-G.; Project administration, F.G.-A. Funding acquisition, C.G.-G.

Funding: This research was funded by the ETSII-Universidad Nacional de Educación a Distancia (UNED) of Spain. Grant grant number 2019/ICF01.

References

1.   Murray, G. La Fundación del Real Ingenio de la moneda de Segovia. In *Real Academia de la Historia y Arte de San Quirce de Segovia 1997 Editores*; Premios Mariano Grau: Segovia, Spain, 1997; pp. 355–542.

2.   Murray, G.S. El Real Ingenio de la Moneda de Segovia: Fábrica industrial más antigua, avanzada y completa que se conserva de la Humanidad. In *Razonamiento Científico de la Propuesta Para su Declaración Como Patrimonio de la Humanidad*; Cámara de Comercio e Industria de Segovia: Segovia, Spain, 2008.

3.   Murray, G.S. *Las Acuñaciones de Moneda en Segovia Desde 30 a.C. Hasta 1869, en Conmemoración de la obra de Rehabilitación del Real Ingenio de la Moneda de Segovia 2007–2011*; Editado por Asociación Amigos de la Casa de la Moneda de Segovia: Segovia, Spain, 2012.

4.   Clarkson, L.A. *Proto Industrialization: The First Phase of Industrialization?* Macmillan International Higher Education: London, UK, 1985.

5.   Alvarez, G.A. *"La Mecanización de la Moneda. El Real Ingenio de Moneda de Segovia, Ejemplo Precoz de Fábrica Industrial", Madrid "Cadenas de Montaje" La Utopía de la Arquitectura Como Producto Industrializado*; II Seminario Internacional G+I_PAI: Madrid, Spain, 2016; pp. 9–31.

6.   Martínez, F.V.S. Estudio Histórico-Tecnológico de las Serrerías de Corte de Piedras Duras en el s. xvi. Aplicación al Análisis y Reconstrucción Gráfica del Molino de Corte de Mármol Utilizado en la Construcción del Retablo Mayor del Monasterio de el Escorial. Ph.D. Thesis, Universidad Politécnica de Madrid, Madrid, Spain, 2016.

7.   Murray, G. *Génesis del Real Ingenio de la Moneda de Segovia: (III) Construcción de los Edificios*; NVMISMA: Madrid, Spain, 1994; núm. 235; pp. 85–119.

8.   Murray, G. *Génesis del Real Ingenio de la Moneda de Segovia: (II) Búsqueda y Concertación del Emplazamiento (1582–1583)*; NVMISMA: Madrid, Spain, 1993; num 232; pp. 177–222.

9.   Murray, G. *Génesis del Real Ingenio de la Moneda de Segovia: (IV) Transporte de la Maquinaria y las Primeras Pruebas (1584–1586)*; NVMISMA: Madrid, Spain, 1994; núm. 235; pp. 85–119.

10.   Parker, G. *El Ejército de Flandes y el Camino Español, 1567–1659: La Logística de la Victoria y Derrota de España en las Guerras de los Países Bajos*; Anaya: Madrid, Spain, 2000.

11.   Fantom, G.S.M.; Izaga, J.M.; Valencia, J.M.S. *El Real Ingenio de la Moneda de Segovia: Maravilla Tecnológica del Siglo XVI*; Fundación Juanelo Turriano: Madrid, Spain, 2006.

12.   Reiner, J.M.I.; Valencia, J.M.S. The Royal Segovia Mint: Hydraulics and devices. In *Renaissance Engineers*; Fundación Juanelo Turriano: Madrid, Spain, 2016; pp. 99–115.

13.   *Ministerio de Obras Públicas, Transportes y Medio Ambiente—Dirección General de Obras Hidráulicas*; Acondicionamiento del Azud de la Casa de la Moneda de Segovia: Segovia, Madrid, 1966.

14.   Tapia, N.G. *Los Veintiún Libros de los Ingenios y Máquinas de JUANELO, Atribuidos a Pedro Juan de Lastanosa*; Departamento de Educación y Cultura: Madrid, Spain, 1997.

15.   Rojas-Sola, J.I.; López-García, R. Engineering graphics and watermills: Ancient technology in Spain. *Renew. Energy* **2007**, *32*, 2019–2033. [CrossRef]

16.   Rudolf, K.F. *Casas de la Moneda Segovia y Hall en Tirol*; Ayuntamiento de Segovia & Instituto histórico Austríaco: Segovia, Spain, 2007; pp. 31–44.

Article

Application of Digital Techniques in Industrial Heritage Areas and Building Efficient Management Models: Some Case Studies in Spain

Carlos J. Pardo Abad [†]

Department of Geography, Universidad Nacional de Educación a Distancia, 28040 Madrid, Spain; cjpardo@geo.uned.es
† Co-director of the "Geography, Landscape and Information Technology" research group.

Received: 19 September 2019; Accepted: 15 October 2019; Published: 18 October 2019

Abstract: This research represents a novel contribution regarding the application of digital technology to the management and cultural promotion of industrial heritage. The study answers questions about the level of digital transformation of certain preselected buildings and areas of great historical and technical interest. It includes an extensive bibliographic review and analyzes different variables linked to webpages, which are the main source of information for visitors, and studies the level of digitization using a survey of the technical managers. The results are valuable because they offer an original profile of selected industrial heritage sites, characterized by an important connection between visitors, visited spaces, and available resources; the interaction of these three elements with the surrounding territory, fostering a new competitive capacity; the projection of each place in a modern and attractive way; and the commitment to an efficient and sustainable local management model. The results provide a fresh look at the technological changes embodied by new uses in old industrialization sites. In addition, the performed analysis could easily be applied and operationally compared in other different heritage environments.

Keywords: digital transformation; industrial heritage; efficient management; ICTs

1. Introduction

The information and knowledge society, internet, e-commerce, connectivity breakthroughs or the generalization of mobile devices and applications have prompted a new technological framework that has greatly impacted all social, economic, and cultural activities. One of the sectors affected most heavily has been tourism, as it aims to adapt to generalized hyperconnectivity and highly interactive tourists.

Visits to former industrial sites are increasingly linked to the expansion of communication channels and the generalization of increasingly sustainable and effective management models. The digital transformation of industrial heritage sites boosts the influx of tourists and their degree of satisfaction which, in turn, encourages the conservation of other historical industrialization sites. This reinforces the role of industrial heritage as an emerging cultural resource of great potential for tourism.

The legacy of industrialization must now be re-read from an integrated, scientific, and innovative perspective: (i) Integrated, because you cannot interpret isolated objects from each other or outside the environment in which they are located, and this implies a territoriality of unquestionable geographical value; (ii) Scientific, because there can be no more purely descriptive approaches to heritage resources, as occurs in a first phase of discovering and cataloguing the industrial heritage, without explanatory causal references; and (iii) Innovative, because a new information and communication technologies-based framework must be built around industrial heritage as a vehicle for the promotion of the destinations, dissemination of associated values, and active interpretation of exhibition contents. At the same time, this threefold aspect serves to ensure resources are managed better.

Through the advanced application of digital techniques, a traditional destination can be turned into a smart destination [1–4]. All these authors consider that these spaces are always linked to technological competitiveness and to improving tourists' experience. Innovative places supported by cutting-edge technological structures that allow for the sustainable development of resources and the integration of visitors with the architectural, environmental, and socioeconomic environment [5].

The application of new technologies in the management of industrial heritage areas and buildings differs from one case to another, but progress, in general, has been positive. New applications and platforms have created great opportunities to bring the legacy of industrialization to the attention of the general public by letting them interact easily and permanently with the resources offered. In this respect, instead of only considering a few issues, like energy efficiency, the sustainability of the retrieved mining and industrial landscapes, the environmental regeneration of the sites, new local employment opportunities, and now, other, more technological aspects linked to information and communication are also taken into account. The spotlight has switched from the place to the visitor, through a multiple sequence of intelligent interpretation of resources.

By creating a complete and effective information system, all involved players can be better coordinated, and goals can be defined more soundly, fostering broader collaboration networks [3,6].

The implications of using ICTs (information and communication technologies) to manage destinations and the strategic, technological, and innovation-oriented vision that presides over new tourism are at the heart of some of the research carried out so far [7–9]. Innovation has been always very significant in relation to competitiveness. It entails processes that affect all players in different contexts and must rely on the development of ICTs to maintain an active interdependence between people and technology. This creates an ecosystem in which multiple human, technical, and economic resources are combined in the form of essential components of tourism intelligence [10].

Cultural tourism could stand to benefit the most from technological advances and the proliferation of new communication tools. Cultural resources are basic for heritage destinations or even complementary for other destinations and, in some cases, heritage preservation is a fundamental precondition for developing the tourist function of the territory and conceptualizing the so-called "territorial heritage system" [11]. Their online promotion and dissemination is an extraordinary opportunity to boost their appeal [12].

Some authors suggest that despite the advantages that ICTs bring by connecting tourists and resources, cultural tourists like to contemplate the original version of the material or intangible item and experience it first-hand. In this regard, new information and communication technologies are not really interpreted as a substitute but, rather, as a complement to the personal experiences recorded during the trip and an instrument to better understand the contents [13,14].

The connection through digital dialogue between institutions, visitors, and objects implies a collaborative approach that increases the opportunities for personal interpretation of resources. This is what underlies a new concept, namely, smart cultural heritage spaces [15,16]. These authors reckon that these spaces can shorten the distance between cultural heritage and visitors and, to a certain extent, overcome the remaining idea that cultural objects are to be enjoyed in a purely aesthetic and essentially passive way.

For all these reasons, new ICTs have unquestionably taken center stage in many of the most recent projects which base their interventions on offering more information, more efficient management, more active visitor participation, and generally reshaping their strategies. It is not simply a matter of replacing heritage resources with digital tools, but of offering more data and participatory services online [17].

Industrial heritage is an emerging cultural resource, although it is still not appreciated enough. The enhancement of its different components has prompted an extensive process of opening up of museums, interpretation centers, cultural parks, ecomuseums, and museum territories [18]. It is essential that all these proposals regarding the exhibition, visualization, and interpretation of industrial heritage be adapted to the powerful existing technological framework. Technological development has created a new cultural reality in industrial heritage. It is a reality defined by ICTs. Digital tools make industrial heritage an intelligent heritage. The most important ones are the following: websites; 3D models; geolocation systems; generation and management of digital representations of physical and functional characteristics of buildings and areas (BIM models); social networks; podcasting; mobile phone apps; QR codes; augmented reality; and multimedia guides.

To date, there has yet to be an in-depth study of the opportunities that new technologies can afford in the industrial heritage field. There are practically no references in this regard, and those that do exist are not very precise and limited to the use of a few general concepts and management models. In some cases, they spotlight industrialization's legacy in the so-called smart territories and the possible advantages of applying innovative techniques [19].

In the field of geography, there are very few investigations related to the new technologies applied to industrial heritage. The studies have basically focused on the analysis of post-industrial landscapes and the transformations experienced from an aesthetic and functional point of view, as well as on the territorial representation of the built elements inherited from industrialization.

As already mentioned, there is some concrete reference to the advantages that the application of technological innovation could have in the management of industrial heritage, all within the framework of the so-called smart territories. It is a geographical vision of undoubted interest that only represents a small approximation to the subject in question.

More specifically, the application of new technologies in the field of heritage education has been addressed, with specific references to industrial heritage [20]. These authors focus on the interpretation of cultural itineraries of industrial heritage, with several case studies in the city of Madrid. The design of a mobile app and its valuation by users becomes the main objective of the research, with an obvious geographical meaning and promotion of the values associated with industrial heritage.

The challenges posed to tourism by the new trends in demand and consumption are based, above all, on the use of new technologies at the service of the best tourist lending. In recent times, the so-called smart destinations have been configured, a new tourism paradigm of deep geographical connotations due to their characteristics of innovative spaces supported by a cutting-edge technological structure, which guarantees the sustainable development of the tourist territory [5].

The particularities of each tourist territory require the implementation of the most appropriate digital solutions [3]. In some cases, the complicated relationships that exist around the large number of local agents involved in tourist destinations (including industrial heritage) are discussed, which makes it difficult to apply digital tools for configuration as smart destinations [21,22]. In any case, the territorial scale of the local type is the most appropriate to promote the digital challenge and overcome the limitations of tourist sites [23]".

Other times, industrial heritage is regarded as a very specific legacy with a great capacity for coordinating territories and resources. The main goal of some authors is to try to define a conceptual model of smart or technological industrial heritage tourism [24]. That task involves developing far more complex forms of management that can be adapted to digital techniques.

These authors define what they call the smart industrial tourism business ecosystem (SITBE) model. It is based on the premise that industrial heritage, considered as a cultural resource, not only calls for investments in the physical recovery of the elements, but also in a range of actions designed to create collaborative organizational structures; involving the local business fabric; boosting technological competitiveness; interacting with the local community; and building a technological model based on intelligent information and communication.

This study mainly aims to ascertain the level of technological innovation at certain industrial heritage sites, chosen on account of their special significance as cultural and tourist resources, and to determine the extent to which digitization enhances efficient management. Another albeit secondary objective is to identify the problems and opportunities found when configuring technically smart industrial heritage destinations.

The lack of specific studies in this field makes it essential to conceptualize what one could refer to as smart industrial heritage spaces; innovative places equipped with technological infrastructure that facilitates permanent access to information, visitor interaction, and sustainable management. Applying digital techniques and platforms in the industrial heritage environment increases the quality of destinations, improves and promotes visitors' active and shared tourist experience, and drives the environment's cultural, environmental and socioeconomic factors as economic development drivers.

2. Method

In this study, we selected eleven different industrial heritage sites, including both isolated constructions and large building complexes, located in the autonomous regions of Andalusia, Castile-La Mancha, Murcia Region, Basque Country, Castile-León, Catalonia, and Asturias. They represent some of the most heavily industrialized sites in Spain's history, which is why it matters so much to ascertain the current level of digital transformation of their management and promotion. In some cases, they are areas with a long-standing mining tradition and a strong initial environmental impact, with interesting constructions and tools of different types [25]. Spectacular landscapes that still bear the deep traces of an activity that lasted a very long time and has resulted in intensive regeneration [26]: Riotinto (Huelva), Almadén (Ciudad Real), La Unión (Murcia), and Samuño (Asturias). The Añana Salt Mines (Alava) are the last example of mining sites selected in the research project.

All other cases involve specific buildings and constructions from different productive sectors adapted to manufacturing techniques. Some date back to the pre-industrial era, like the San Blas Ironworks (León), the Royal Glass Factory in La Granja de San Ildefonso (Segovia), and the Royal Spanish Mint in the city of Segovia [27,28]. Others date back to the industrial era per se, such as the La Encartada Beret Factory in Balmaseda (Vizcaya), the water lift in Cornellà del Llobregat (Barcelona), and the Vapor Aymerich, Amat y Jover textile factory in Tarrasa (Barcelona) (Figure 1).

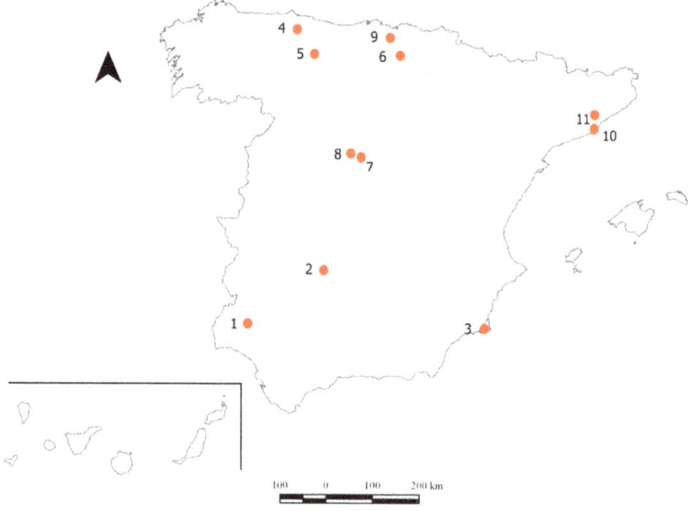

Figure 1. Location of buildings and areas analyzed: 1. Riotinto mining area; 2.Almadén mining area; 3. La Unión mining area; 4. Samuño mining area; 5. San Blas Ironworks; 6. Añana Salt Mines; 7. Royal Glass Factory; 8. Royal Spanish Mint; 9. La Encartada Beret Factory; 10. Water lift; 11. Vapor Aymerich, Amat y Jover textile factory. Source: José Fernández Álvarez (UNED SIG Laboratory).

From a methodological viewpoint, an extensive review of the scientific literature has been conducted to verify the different theoretical approaches present in the smart tourism field, the application of new digital technologies, and the proposed innovative resource management models. Also with regard to the method, one proposal put forward is to analyze them by using different variables present on their webpages, creating an index and conducting a survey.

Use of the web popularity index (WPI) is proposed to check and measure the variables regarding the content and information available on the webpages. The WPI, which ranges from 0 to 1, measures the percentage compliance of a broad range of selected variables that each industrial heritage site had in February 2019. The results are analyzed both in terms of constructions and sites and of variables; in other words, identifying the mean value obtained in each building and site for all variables and of each variable for all case studies.

The WPI is a very interesting tool in this investigation. It is considered a fundamental part of the study not previously used by other researchers. Obtaining the indicator is quick and simple, and the results are very direct and understandable from the point of view of the technological reality used in the case studies. The WPI quantifies different and highly significant qualitative aspects and is very useful in two ways: first of all, it is an efficient, straightforward, and simple way of approximating the information facilitated to visitors; and, secondly, it can be used to categorize the sites and buildings selected in terms of their level of application and display in digital environments.

A key part of the method has been the survey, carried out between the months of March and April 2019, among the technical managers and directors of the selected industrial heritage sites. The blocks and questions are listed in Table 1 below, and the levels of response were limited to three categories: agree, disagree, and no opinion. The survey, divided into five blocks and thirty different questions with one final open-ended question, has provided digital reality information about the management, sustainability, accessibility, connectivity, innovation, and use of ICTs.

Table 1. Blocks and questions in the model survey conducted.

Block	Question
(A) General solutions and trends for creating a smart building or site	1. The current technological scenario offers interesting opportunities in this respect 2. Implementing technology is regarded as a renewal and revaluation measure in any case 3. New technologies favor greater interaction between visitors, buildings, or sites 4. Applying technology is perceived as a good management move 5. Applying new technologies is complex and sometimes problematic 6. When applying technologies, the economic cost is considered a decisive factor
(B) Specific solutions and trends for the industrial heritage building or site management	7. Technology is already playing a key role in the management and promotion of your industrial heritage building or site 8. Using new technologies is considered to provide greater efficiency and reduces costs to a great extent 9. The members and workers of the industrial heritage building or site are aware of and properly trained in new digital technologies 10. There is a suitable level of collaboration between different (public and private) local and regional entities for implementing technology in the industrial heritage building or site 11. The municipal authorities help to promote the site or building as a smart space 12. The industrial heritage building or site has a specific strategy for becoming a smart space 13. The industrial heritage building or site is evolving favorably towards being considered a smart space

Table 1. *Cont.*

Block	Question
(C) Specific solutions and trends for the industrial heritage building or site: sustainability and accessibility	14. The industrial heritage building or site is sustainable in relation to the conservation of the surrounding landscape and the available resources 15. The industrial heritage building or site applies light energy saving by using led technology 16. The industrial heritage building or site has lighting control sensors based on luminosity conditions 17. The industrial heritage building or site applies water saving measures 18. The zone is accessible for disabled people 19. The industrial heritage building or site has tactile models that can be interpreted by the visually impaired 20. Digital access to all the information about the industrial heritage building or site is promoted
(D) Specific solutions and trends for the industrial heritage building or site: connectivity and innovation	21. Internet connectivity is adequate 22. You can log onto free Wi-Fi networks 23. The industrial heritage building or site is innovative regarding the inclusion of new technologies 24. The industrial heritage building or site has a tourist application for smartphones 25. The industrial heritage building or site has interpretive panels with QR codes 26. The QR codes are included in the printed promotional material
(E) Specific solutions and trends for the industrial heritage building or site: application of information and communication technologies (ICTs)	27. These technologies are used to know more about registered visits 28. ICTs are used to ensure better management of the resources of the industrial heritage building or site 29. ICTs are used in the promotion of the industrial heritage building or site 30. ICTs are used to improve visitors' experience in the industrial heritage building or site

Source: own elaboration.

3. Results

The internet is one of the biggest information access windows, and instantly provides knowledge to a large part of the world's population. This technology is essential for making industrial heritage more accessible and making the historic heritage of industrialization more visible.

Tourists have become more and more interested in ICTs over the latest decades and it is calculated that right now, around 45% of internet searches about tourism in Spain, similarly to all other developed countries, are related to cultural aspects.

Webpages are the main source of information that people look at when planning a trip, so a building or place stands to gain a lot in information and value terms if it has an attractive, well-designed website. The fact that industrial heritage appeals to a selective, minority kind of tourism reinforces the idea that it is essential to have a technological scenario appropriate to the digital tourist's needs and to "democratize" online knowledge of industrialization's heritage and the final success of new-use projects.

3.1. Webpages and Digital Display of Content

In order to analyze the webpages of the cases selected for the study, first, a total of 14 different variables regarding the digital information available to tourist were established. Those variables are (i) attractive design and well-structured content; (ii) presentation of interesting content; (iii) smart event calendar log; (iv) availability of weather information; (v) access to multilingual information; (vi) possibility of booking tickets online; (vii) online promotion of other representative tourist attractions; (viii) destination displayed with Google Maps™ and/or Street View; (ix) possibility of using mobile device applications; (x) integrated presence in social media (Facebook, Twitter, YouTube, etc.); (xi) possibility of giving opinions and ratings; (xii) availability of a virtual tour; (xiii) access to PDF digital leaflets in or 3D models from photographs; and (xiv) access to multimedia tools (videos, audios, augmented reality, etc.).

The heritage site with the highest web popularity index (WPI) is the Almadén mining area (now the Almadén Mining Park) (Figure 2). The next buildings or areas, with a slightly lower WPI, are as follows: Vapor Aymerich, Amat y Jover textile factory (now the Catalonia National Museum of Science and Technology, MNACTEC), water lift (now the AGBAR Water Museum), Royal Glass Factory (now the Glass Technology Museum), Real Casa de la Moneta, Añana Salt Mines and San Blas Ironworks (now the Castile-León Iron & Steel and Mining Museum). Almost all the specified variables are present in this first level of selected buildings and areas, which can be regarded as spaces with a highly technological web presence. Their webpages offer a varied range of information and digital resources of different kinds, in line with the expectations generated as sites that are highly symbolic as far as industrial heritage is concerned. In all these cases, the web popularity index is equal to or higher than 0.78.

Figure 2. Almadén mining area. Author: Carlos J. Pardo Abad.

On a second level are all the other analyzed cases with lower scores than the first group, between 0.78 and 0.43, namely, the Riotinto, La Unión, and Samuño mining areas (all three of which have been turned into mining parks and receive a large number of visitors every year). The La Encartada Beret Factory ranks last among all the cases studied with a WPI of 0.43. According to these figures, each selected building and site ranks in different positions, depending on the level of technology applied on their webpages (Figure 3).

The general average of the complied variables is 70%, which represents a good level as a whole, but could be improved in certain aspects. The variable with the highest level of compliance refers to the presentation of interesting content (100%), followed by the attractive design and well-structured content (91%), smart event calendar log (91%), online promotion of other representative tourist attractions (91%), and integrated presence in social media (91%). The access to multilingual information, destination displayed with Google Maps™ and/or Street View, and access to PDF digital leaflets or photograph-based 3D model variables were complied with in 82% of cases.

In the other variables, the compliance percentages ranged from 73% to 1%. Weather information was only available in one of the selected cases, which is quite surprising as weather conditions affect the number of people visiting industrial heritage buildings and sites. Levels of presence are also low in the variables regarding the possibility of using mobile device apps (18%) and the availability of a virtual tour (36%). The possibility of expressing opinions and giving scores online is clearly above average (64%), while compliance with the other variables is around 70% (Table 2).

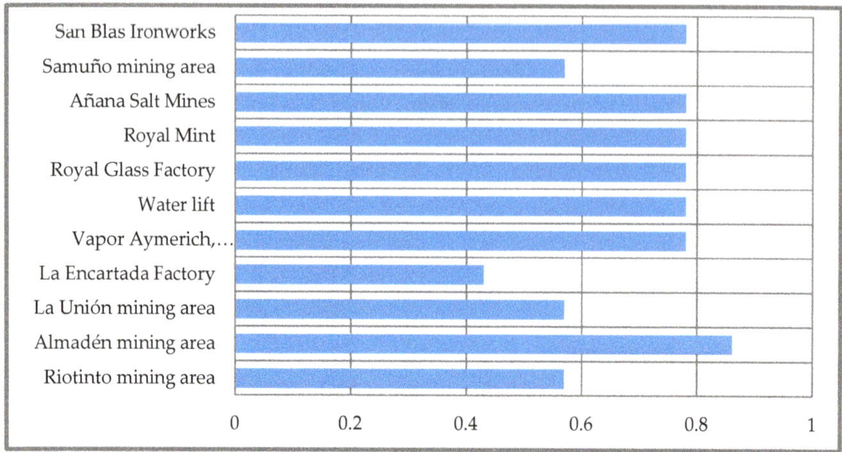

Figure 3. Web popularity index (WPI) of industrial heritage buildings or areas. Source: own elaboration.

Table 2. Overall level of compliance of the selected variables.

Variable	Percentage
1. Attractive design and well-structured content	91
2. Presentation of interesting content	100
3. Smart event calendar log	91
4. Availability of weather information	1
5. Access to multilingual information	82
6. Possibility of booking tickets online	73
7. Online promotion of other representative tourist attractions	91
8. Destination displayed with Google Maps™ and/or Street View	82
9. Possibility of using mobile device applications	18
10. Integrated presence in social media (Facebook, Twitter, YouTube, etc.)	91
11. Possibility of giving opinions and ratings	64
12. Availability of a virtual tour	36
13. Access to PDF digital leaflets in or 3D models from photographs	82
14. Access to multimedia tools (videos, audios, augmented reality, 360° panoramas, etc.)	73
Mean value ($\overline{\mathbf{X}}$)	70

Source: own elaboration.

Multilingual information can be found at all the selected sites, except in the La Unión and Samuño mining areas. In the other case studies, information is always offered in Spanish and English and, more specifically and occasionally, in other languages. On the issue of this variable, it is striking that on the Riotinto area website (Figures 4–6)—corresponding to an area in the province of Huelva close to the Portuguese border and which attracts numerous visitors from the neighboring country—the information is not available in Portuguese.

Figure 4. Orthophoto and location of the Riotinto mining area. Source: José Fernández Álvarez (UNED SIG Laboratory).

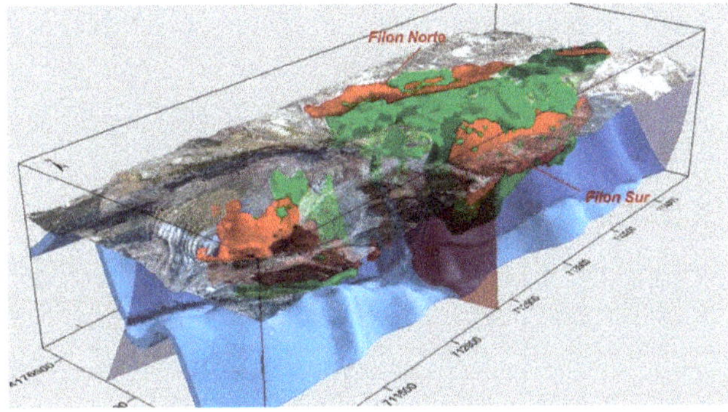

Figure 5. 3D model of the Riotinto mining area [29]. This figure shows an axonometry, from the SE, of the open pit mines of Cerro Colorado and Corta Atalaya. In blue, the anticline bounded by the northern and southern faults. The zones of copper mineralization appear in green and the pyritic stockwork appear in red.

Figure 6. Riotinto mining area. Author: Carlos J. Pardo Abad.

3.2. Survey and Inference of the Level of Technological Management

The survey used was sent to the technical managers of the selected heritage sites and constructions, with several sets of questions on management, sustainability, accessibility, connectivity, innovation, and ICT usage. The level of response was 64%, meaning that the results are reliable. The survey was divided into five blocks, to which a final open-ended question was added. The first block (Block A) was of a general nature, and was entitled "General solutions and trends for creating a smart building or place". The next ones refer exclusively to the level of technological intelligence applied to each selected industrial heritage site: the second block (Block B) is on "management", the third (Block C) on "sustainability and accessibility", the fourth (Block D) on "connectivity and innovation", and the fifth (Block E) on "application of information and communication technologies". These blocks can be seen in Table 3.

Table 3. Summary of the levels of response by survey blocks.

Blocks	Thematic Area	Agree (%)	Disagree (%)	No opinion (%)
Block A	General	83	14	3
Block B	Management	50	40	10
Block C	Sustainability and accessibility	73	22	5
Block D	Connectivity and innovation	50	48	2
Block E	ICT	61	21	18
Total		63	29	8

Source: own elaboration. Note: highlighted fields have the highest response percentages.

Table 3 shows a general summary of the survey answers, with average percentages for all the questions contained in each block. There is a majority of "Agree" answers in all cases, especially Block A, with 83% as the highest score, and Blocks B and D, with 50% as the lowest scores. Therefore, it is these last two blocks that have the highest percentages of "Disagree" answers, with 40% and 48%, respectively.

Worth highlighting about the results of the first block of the survey (Block A) is that all the site managers agreed that the current technological scenario offers interesting opportunities for building a smart space. They also appreciate that technologies foster greater interaction with visitors.

There were bigger differences in the other questions of this block because 83% agreed that implementing technology is a means of renewing and reasserting the value of the site, as well as improving their management, while the remaining 17% disagreed. The same response percentages were obtained in the question as to whether economic cost is considered a decisive factor when applying technologies. In the last question of this first block, 50% of the respondents agreed with the idea that applying new technologies is complicated.

As explained earlier, the following four blocks of questions refer to the specific heritage building or area. On the management issue (Block B), 50% of respondents saw technological solutions and trends as a good option for managing industrial heritage, although a not too insignificant percentage, 40%, disagreed. The remaining 10% had no opinion.

Asked about whether technology already plays a fundamental role in management and promotion, 50% of the sample agreed and the other 50% disagreed. There were greater imbalances in the answers to the other questions. Around 83% agreed with the fact that using the new technologies affords greater efficiency and cuts costs. Around 66% considered that the industrial heritage building or area's members and workers are sensitized to and properly trained in new digital technologies, and 83% of respondents believed that local and regional entities do not cooperate enough in implementing technology, but did acknowledge the role played exclusively by municipal authorities.

The surveys show that 50% of the museums open in the industrial heritage building or sites already have a specific strategy for becoming a smart tourist destination. This positive situation is linked directly to the idea that the building or place is evolving favorably towards being considered a smart space in 66% of cases. In any case, it seems that all the measures devised to further promote these strategies in the short and medium term need to be increased.

The next block of the survey (Block C) has to do with two basic issues of great importance for the promotion of the preserved assets: sustainability and accessibility. It was found that 73% of the surveyed places already apply sustainable measures, such as energy and water saving, aesthetic conservation of the landscape, and maintenance of available heritage resources. Also worth noting are the site accessibility-related measures. Only 22% disagreed and the remaining 5% had no opinion.

In all cases, respondents considered that the industrial heritage building or site is already sustainable in relation to the landscape and available resources. This level of response represents full awareness regarding the aesthetic value of the environment in this type of heritage (especially notable in mining areas). The landscape and architectural or technical resources are always stimulating factors in the cultural promotion of the places and as spaces that are locallybased identity reference points.

The levels of responses about energy or water saving measures differed more. In most cases, they are already applied in lighting (86%) with the use of LED technology, and in 14% of cases, there are sensors for regulating lighting in terms of light conditions. Water saving measures are in place in 72% of cases.

Physical and digital accessibility is another important aspect in this block. According to 86% of respondents, the industrial heritage building or area is accessible to people with physical disabilities. As regards visual disability, only 57% of the sites have tactile models that can be interpreted by the blind or visually impaired; and when asked about digital accessibility, all the selected places said that they promote web access to all their information.

The next block of questions (Block D), penultimate of the whole survey, refers to connectivity and innovation. Broadly speaking, 50% of respondents considered that the level of application is not yet high enough, while 48% agreed and the remaining 2% had no opinion.

Internet connectivity is adequate at the destination in 86% of cases, but free Wi-Fi networks are only available at 28% of the sites. This result is especially striking because it is one of the technological measures that is easiest to apply. Similar figures (28% agree) were obtained when asked about the availability of a tourist app for smartphones (Figures 7 and 8).

Figure 7. Vapor Aymerich, Amat y Jover (MNACTEC). Author: Carlos J. Pardo Abad.

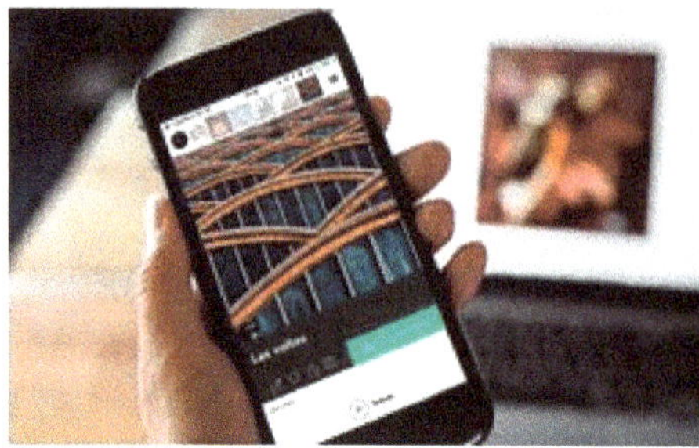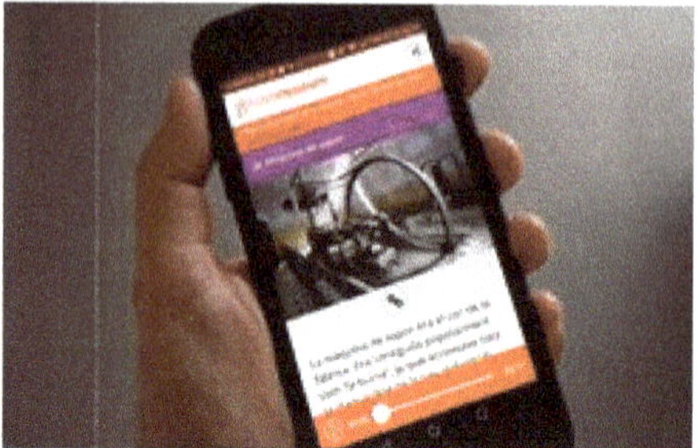

Figure 8. Several MNACTEC applications for smartphones. Left: *Cloudguide* App (virtual tour of the museum's objects and spaces); right: *Visitmuseum* App (audio guide of the Vapor Aymerich, Amat and Jover building). Source: MNACTEC.

The survey shows that those responsible for the destinations are aware that much remains to be done in terms of bringing in new technologies, despite the recognized advantages that these solutions afford in terms of more efficient and sustainable resource management. In 43% of cases, respondents considered that the sites are already innovative, but the very same percentage thought that they are not yet.

Around 57% of the sites have interpretive panels with QR codes and in 43% of cases, these codes are thought to already be included in some type of printed promotional material. These are still low percentages that should increase in the coming years out of necessity.

The last block of the survey (Block E) refers to the application of information and communication technologies. They are used in 61% of cases and, in addition, in 57% they are used both to know more about tourism demand and to improve the visitors' experience and manage existing resources more effectively. Broadly speaking, it can be said that ICTs are used more for promotion than for management, because 72% of technical managers agree on their validity for publicizing the values associated with the preserved elements of industrial heritage.

Table 4 summarizes the highest levels of acceptance by blocks and questions, and according to each industrial heritage building or area.

Table 4. Summary of the highest levels of acceptance per industrial heritage building or area, according to blocks and questions.

Building or Area	Blocks and Questions
Riotinto mining area	A1–6, B7–9, B12–13, C14–20, D21, D23, E27–30
Almadén mining area	NC
La Unión mining area	A2–6, B7–9, B13, C14–15, C20, D21, D24–26
Samuño mining area	NC
San Blas Ironworks	A1–4, A6, B8–9, B13, C14, C18–20, D21, D26, E29
Añana Salt Mines	A1–4, A6, B7, B9, B13, C14–15, C17–20, D23, E27–30
Royal Glass Factory	A1–6, B8, B12, C14–15, C17–20, D21–22, D25–26, E28–30
Royal Spanish Mint	NC
La Encartada Beret Factory	A1, A,3, A4, A6, B7–10, B12–13, C14–15, C17–18, C20, D21, D23, D25, E27–30
Water lift	NC
Vapor Aymerich, Amat y Jover textile factory	A1–3, A5, C14–15, C17–18, C20, D21–22, D24–25, E27

Source: own elaboration. Note: A1 (Block A, question one; B2–4 (Block B, questions 2 to 4), etc. NC: no answer.

The last survey question, as explained above, was open-ended. Only in one case has there been an answer, specifically from the managers of the Riotinto mining area. The answer indicated that right now, they are actively working on steadily implementing new digital technologies. Another significant item of information they added was that in just one year after implementing the electronic ticketing system, the percentage of visitors to the mining park has risen by 45%. This clearly points to the role that digital technologies can play in cultural resource management and promotion, in this case, in relation to the heritage linked to the industrialization and production techniques of times gone by.

4. Discussion of Results and Conclusions

This study created a model for analyzing the existing digital level in a series of industrial heritage buildings and areas, selected for their special significance in the conservation of built structures and industrialization landscapes, and for their current use as culture sites. This article represents a unique approach to a thematic area that is under-researched, perhaps because industrial heritage tourism is a very specific market and its ties to the latest innovation models are still undervalued.

The advances made in the digital ICT field have created previously unknown connections between visitors, visited spaces, and available resources. The interaction of these three aspects among themselves, and with the neighboring territory, is fostering a set of new-generation competitive capabilities that enhance the projected image of each place and consumer demand for these cultural goods.

Digital techniques are part of the more general methodological process of what is known as "territorial intelligence". This intelligence emphasizes geospatial databases, the creation of territorial observation laboratories, the characterization of all the players involved or the application of collaborative models between entities, managers, and visitors [3]. It is a global strategy that constitutes a kind of new approach in the most recent analysis of digitally innovative places, which involve several non-exclusively technological elements [23,30].

The most efficient and sustainable management models represent the transition from theory to practice, always necessary in an era of great technical opportunities and growing digital competitiveness. In industrial heritage buildings and areas, these models are particularly significant due to the very characteristics of the preserved assets and the profound environmental changes caused by decades of prolonged economic activity [26].Some geographical research is of great importance because it constitutes an empirical study on some specific case or subject related to new technologies in the field of industrial heritage [20]. This supposes a qualitative leap that surpasses reflections of a strictly theoretical nature and a relevant scientific contribution on the validity of new technologies in tourism management and promotion of industrial heritage.

The results achieved in this research—almost the first of its kind in the field of industrial heritage-related technological innovation—are evident and exceed any general and exclusively theoretical approach. The results in themselves represent an original perspective of the current context of technological change and an approximation to the human and technical efforts made in the museums and cultural centers installed in the material elements of industrialization. Another outstanding feature of the analysis is that it is easy to apply and operationally compared in other different environments.

The advantage of the selected places is that they became tourist attractions when the digital revolution started up, and so have always invested in technological solutions as a means to avoid being identified as a traditional destination. Many a difficulty has arisen applying innovations, sometimes due to a coordination between the different players involved, and others due to the strategic lack of cooperation networks between the different industrial heritage sites, the high front-end costs, and the technical complexity perceived in the early stages.

There are several parts to the analysis conducted during this research: first, a critical and reasoned review of existing scientific literature in this regard; and, next, the selection of the case studies, an in-depth examination of their webpages, and the completion of a survey by the technical managers and directors.

Digital intelligence is generally perceived as a good management reference and an excellent scenario for interacting with visitors. The levels of innovation found are not uniform, as occurs with other kinds of heritage, with their own features adapted to the surrounding territory and to the internal logic of each production process. This diversity inevitably delays any homogeneity in applying any digital innovation.

The webpages have been studied using a broad set of qualitative variables and producing a web popularity index (WPI) that simply and efficiently summarizes the information available in each place. The case study with the highest WPI is the Almadén mining area. The fact that its varied digital content is so carefully presented could have to do with it having been listed, since 2012, as a World Heritage Site by UNESCO, due to its exceptional values associated with the mining and international mercury sales. The next most outstanding examples, in terms of general compliance with the variables, are the San Blas de Sabero Ironworks, Añana Salt Mines, Royal Segovia Mint, Royal Glass Factory of La Granja de San Ildefonso, Cornellà de Llobregat water lift, and Tarrasa Vapor Aymerich, Amat y Jover textile factory. All these case studies show above-average compliance of the variables established as an analysis reference. In the rest, the digital level drops and the information provided on the website is less complete.

The variable with the highest level of compliance refers to the presentation of interesting content, followed by the attractive design and well-structured content, smart event calendar log, online promotion of other representative tourist attractions, and integrated presence in social media.

The survey shows that the majority idea is that when applying technology, a decisive role is played by the economic cost, as well as the problems ensuing from its complexity. In any case, the majority opinion is that the selected places are evolving favorably towards being considered smart destinations (66%).

The lowest scoring of all the blocks into which the survey was divided is the connectivity and innovation block, followed by the management block. In contrast, the sustainability and accessibility block is quite favorably perceived, especially regarding conservation of the landscape and available resources, the application of energy and water saving measures, and physical and digital accessibility to information. Notably significant in terms of the application of ICTs is their use in promoting tourism, and less in terms of ascertaining demand, resource management, and improving visitors' personal experience. Compensating for this imbalance is essential to improving the levels of digital intelligence and efficient management in the medium term, as well as to forecast the emergence of new consumer models with sounder criteria.

Several main objectives must be achieved in the coming years to boost technological innovation and overcome current difficulties. An objective will be the best coordination between the different public administrations and local agents. Another objective will be the strategic creation of collaboration networks between the different places of industrial heritage. It will also be important to seek for funding to reduce high technological costs in the early stages of innovation, and specialized technical assistance that reduces the complexity of the implementation and development of new technologies. With these measures and other complementary ones, the digital tools will expand the cultural and tourist opportunities. The measures will facilitate the management of industrial heritage sites and will serve to disseminate the values of a legacy of enormous scientific and technical interest, with appropriate digital media for each resource.

The process of technological improvement is slow and continuous over time, that is, in a permanent phase of change to introduce the necessary advances for the most active interaction with the cultural user in the selected places of industrial heritage. In some analyzed museum, the active effort for the progressive application of new technologies is recognized. As an example, the implementation of the electronic sales system has increased the final visits and the percentage of sales with this technical means. They are very important and significant advances for the near future and for the general recognition of a legacy of great cultural importance.

The spaces studied are all local ones. This represents an advantage for achieving the innovation and efficiency goals because the new management methods and tools have a faster impact in smaller spaces due to their lower environmental and constructive complexity. Digital advances have been important, but there is still a need to keep on extending the use of new technologies and interconnecting all technical and human components for the same purpose. That is the only way to build a smart ecosystem capable of successfully promoting the image of industrial heritage as a resource for cultural consumption.

Funding: This research has been funded by the Research Challenges R&D&i Project entitled "Vulnerability, resilience and strategies for the reuse of heritage in deindustrialized spaces". Ministry of Science, Innovation and Universities, 2018 call. Reference: RTI2018-095014-B-I00. Lead Researcher: Paz Benito del Pozo, University of León.

Acknowledgments: I would like to express my thanks for the support offered by the technical managers of the places selected in the research, without which it would have been impossible to carry out the study. I am also grateful for the cooperation of José Fernández, at the UNED SIG Laboratory, for producing some of the figures included in the text.

Conflicts of Interest: The author declares no conflict of interest.

References

1. Lamsfus, C.; Alzur-Sorzabal, A. Theoretical framework for a tourism internet of things: Smart destinations. *J. Tour. Hum. Mobil.* **2013**, *2*, 15–21.

2. Buhalis, D.; Amaranggana, A. Smart tourism destinations. In *Information and Communication Technologies in Tourism 2014*; Xian, Z., Tussyadiah, L., Eds.; Springer: Irelnd, UK, 2013; pp. 553–564.

3. Luque, A.M.; Zayas, B.; Caro, J.L. Los destinos turísticos inteligentes en el marco de la inteligencia territorial: Conflictos y oportunidades. *Investig. Tur.* **2015**, *10*, 1–25. [CrossRef]

4. Boes, K.; Buhalis, D.; Inversini, A. Smart tourism destinations: Ecosystems for tourism destination competitiveness. *Int. J. Tour. Cities* **2016**, *2*, 108–124. [CrossRef]

5. López de Ávila, A.; García, S. Destinos turísticos inteligentes. *Harv. Deusto Bus. Rev.* **2013**, *224*, 58–67.

6. Miralbell, O.; Sivera, S. New innovation networks in destinations 2.0. In *XVI Simposi Internacional de Turisme i d'Oci*; ESADE: Barcelona, Spain, 2008.

7. Plaza, A.G.; Herrero, J.L.C.; Maldonado, A.A.; Rossi, C.; Olivencia, J.L.L. Sistema integrado de gestión de destinos. In *VIII Congreso Sobre Turismo y Tecnologías de la Información y Las Comunicaciones TURITEC 2010*; Universidad de Málaga: Málaga, Spain, 2010.

8. Buhalis, D.; Law, R. Progress in information technology and tourism management: 20 years on and 10 years after the Internet. The state of e-tourism research. *Tour. Manag.* **2008**, *29*, 609–623. [CrossRef]

9. Volo, S. Bloggers' reported tourist experiences: Their utility as a tourism data source and their effect on prospective tourists. *J. Vacat. Mark.* **2010**, *16*, 297–311. [CrossRef]

10. Gretzel, U.; Werthner, H.; Koo, C.; Lamsfus, C. Conceptual foundations for understanding smart tourism ecosystems. *Comput. Hum. Behav.* **2015**, *50*, 558–563. [CrossRef]

11. Troitiño, M.A.; Troitiño, L. Patrimonio y turismo: Reflexión teórico-conceptual y una propuesta metodológica integradora aplicada al municipio de Carmona (Sevilla, España). *Scr. Nova. Rev. Electrón. Geogr. Cienc. Soc.* **2016**, *543*, 1–45.

12. Suau, F. El turista 2.0 como receptor de la información turística: Estrategias lingüísticas e importancia de su estudio. *Pasos. Rev. Tur. Patrim. Cult.* **2012**, *10*, 143–153. [CrossRef]

13. Rojas-Sola, J.I. Patrimonio cultural y tecnologías de la información: Propuestas de mejora para los museos de ciencia y tecnología y centros interactivos de Venezuela. *Interciencia* **2006**, *31*, 664–670.

14. Guttentag, D.A. Virtual reality: Applications and implications for tourism. *Tour. Manag.* **2010**, *31*, 637–651. [CrossRef]

15. Chianese, A.; Piccialli, F. SmaCH: A framework for smart cultural heritage spaces. In Proceedings of the Tenth International Conference on Signal-Image Technology & Internet-Based Systems, Marrakech, Morocco, 23–27 November 2014.

16. Chianese, A.; Piccalli, F.; Valente, I. Smart environments and cultural heritage: A novel approach to create intelligent cultural spaces. *J. Locat. Based Serv.* **2015**, *9*, 209–234. [CrossRef]

17. Gómez, A.; Server, M.; Jara, A.J. Turismo inteligente y patrimonio cultural: Un sector a explorar en el desarrollo de las smart cities. *Int. J. Sci. Manag. Tour.* **2017**, *3*, 389–411.

18. Pardo, C.J. Sostenibilidad y turismo en los paisajes culturales de la industrialización. *Arbor. Cienc. Pensam. Cult.* **2017**, *193*, 400. [CrossRef]

19. Álvarez, M.A. La herencia industrial y cultural en el paisaje: Patrimonio industrial, paisaje y territorios inteligentes. *Labor Eng.* **2010**, *4*, 78–100. [CrossRef]

20. De la Peña, F.D.; Hidalgo, C.; Palacios, A.J. Las nuevas tecnologías y la educación en el ámbito del patrimonio cultural. <<Madrid Industrial, Itinerarios>>. Un ejemplo de m-learning aplicado al patrimonio industrial. *Tecnol. Cienc. Educ.* **2015**, *2*, 51–82.
21. Buhalis, D. Marketing the competitive destination of the future. *Tour. Manag.* **2000**, *21*, 97–116. [CrossRef]
22. Aldebert, B.; Dang, R.J.; Longhi, C. Innovation in the tourism industry: The case of tourism@. *Tour. Manag.* **2011**, *32*, 204–213. [CrossRef]
23. Ivars, J.A.; Solsona, F.J.; Giner, D. Gestión turística y tecnologías de la información y la comunicación (TIC): El nuevo enfoque de los destinos inteligentes. *Doc. Anàl. Geogr.* **2016**, *62*, 327–346.
24. Perfetto, M.C.; Vargas-Sánchez, A.; Presenza, A. Managing a complex adaptive ecosystem: Towards a smart management of industrial heritage tourism. *J. Spat. Organ. Dyn.* **2016**, *4*, 243–264.
25. Pardo, C.J. The post-industrial landscapes of Riotinto and Almadén, Spain: Scenic value, heritage and sustainable tourism. *J. Herit. Tour.* **2017**, *12*, 331–346. [CrossRef]
26. Pardo, C.J. Environmental Recovery of Abandoned Mining Areas in Spain: Sustainability and New Landscapes in Some Case Studies. *J. Sustain. Res.* **2019**, *1*, 190003.
27. Pardo, C.J. *El Patrimonio Industrial en España. Paisajes, Lugares y Elementos Singulares*; Editorial Akal: Madrid, Spain, 2016; p. 286.
28. Pardo, C.J.; Martínez, J. Conservation, management and tourist use of pre-industrial heritage. Identification of Spanish experiences from a territorial analysis. *J. Tour. Hosp. Manag.* **2015**, *3*, 1–22.
29. Martín-Izard, A. A new 3D geological model and interpretation of structural evolution of the world-class Río Tinto VMS deposit, Iberian Pyrite Belt (Spain). *Ore Geol. Rev.* **2015**, *71*, 457–476. [CrossRef]
30. Caro, J.L.; Luque, A.; Zayas, B. Nuevas tecnologías para la interpretación y promoción de los recursos turísticos culturales. *Pasos. Rev. Tur. Patrim. Cult.* **2015**, *13*, 931–945.

applied
sciences

Article

Academic Proposal for Heritage Intervention in a BIM Environment for a 19th Century Flour Factory

Alberto Sánchez [1,*], Cristina Gonzalez-Gaya [2], Patricia Zulueta [1], Zita Sampaio [3] and Beatriz Torre [1]

1   Department of Materials Science and Metallurgical Engineering, Graphic Expression in Engineering, Cartographic Engineering, Geodesy and Photogrammetry, Mechanical Engineering and Manufacturing Engineering, School of Industrial Engineering, Universidad de Valladolid, Paseo del Cauce 59, 47011 Valladolid, Spain
2   Department of Construction and Manufacturing Engineering, ETSII-Universidad Nacional de Educación a Distancia (UNED), C/Juan del Rosal 12, 28040 Madrid, Spain
3   Department of Civil Engineering, Architecture and Georesources, Instituto Superior Técnico, Universidade de Lisboa, Avenida Rovisco Pais, 1049-001 Lisboa, Portugal
*   Correspondence: asanchez@eii.uva.es

Received: 19 August 2019; Accepted: 29 September 2019; Published: 2 October 2019

Abstract: The implementation of building information modeling (BIM) has become a reality worldwide, not only because of the advantages it offers, but also because of the obligatory nature of its use in construction and civil engineering projects in various countries around the world. An intervention project on an industrial heritage property requires a methodology that considers the condition of the building over time and its value for new use. The advantages of working with a precise 3D model that integrates engineering data in a collaborative work environment makes BIM and heritage BIM (HBIM) very useful tools in a project whose objective is the recovery of an industrial heritage real estate property. This work is part of the academic implementation of BIM in university technical education centers and aims to establish a methodology for shared and collaborative group work in a BIM environment through a Spanish industrial heritage case study of a flour factory dating to 1865. A rigorous historical study and the elaboration of a central BIM model loaded with real content on the industrial complex have allowed the immersion of the students into the BIM methodology, as well as the generation of a value proposition for the exploitation of the factory.

Keywords: BIM; collaborative methodologies; industrial heritage; HBIM

---

1. Introduction

The difference between existing common buildings and heritage buildings lies in the fact that the latter involve the architectural, historical, and archeological documentation necessary for the technical reproduction of a given context, as well as an intellectual effort to describe the cultural-leisure context [1].

Various theorists of art and architecture, which were important figures of knowledge of the nineteenth and twentieth centuries, have proposed many theories, which are not always aligned, on the restoration of heritage buildings. Table 1 shows some of the most relevant.

In Spain, during the first quarter of the twentieth century, ancient monuments were restored by professionals who mostly followed the methods of Viollet-Le-Duc. Parts of the monuments that, according to the criteria of the restorer, did not belong to the original style were replaced by copies or interpretations of the originals. During restoration, the Spanish monuments lost the effects of time, and most were completely rebuilt [2].

The Spanish architect and restorer Leopoldo Torres Balbás (1888–1960), who began work as a curator of the Alhambra and the Generalife of Granada in 1923, recognized the reality of restoration occurring in Spain and developed a new innovative system of intervention in historical architecture [2]. Beginning in 1931, through his participation in the Athens Conference, he contributed to elevating the precepts formulated there to a normative position. Thus, the Athens Charter became the first manifesto of the theory of architectural heritage restoration.

Industrial heritage assets must be jointly understood as assets and as having an industrial nature. For the former, they must be carriers of sufficient value and have the ability to transmit it; industrial character is defined by a productive activity that motivates the existence of the good. Within the material assets of industrial heritage, we can distinguish between movable assets (the machinery that is part of the production process) and real estate (the factory as a productive space), which are both linked [3].

To address new challenges of the twenty-first century in relation to industrial heritage, as a result of the VII Seminar on Industrial Landscapes of Andalusia, the 2018 Seville Charter of Industrial Heritage was published in 2019 and signed by more than twenty experts from different scientific disciplines. Its purpose was to delimit the different approaches that affect industrial heritage so that advancement of their knowledge is possible and coordinated and comprehensive strategies could be proposed to respond to the issues arising from their maintenance and conservation. The document provides a series of recommendations aimed at enhancing the historical legacy of industrial heritage [4].

On the one hand, it is necessary to preserve a site and its historical or cultural value as part of the identity of a community to study the past and best understand the present; on the other hand, it is necessary to provide that territory with a new meaning that does not hide or minimize the original, but that provides an added function that will continue to be valued in a positive context, serving its use for the development of the environment in which it is located and, if necessary, for its regeneration [5].

Table 1. Theories for the restoration of heritage assets [2–12].

Author	Theory
Viollet-Le-Duc (1814–1879)	The building should retain the style of the period in which it was built. All architectural elements from a different era should be eliminated, thus also eliminating its historical stratification; nondifferentiated lasting intervention of the original historic building.
John Ruskin (1819–1900)	Build a new reality on the remains of a pre-existing building. Provide sound maintenance to apply the necessary operations for conservation. Preserve with the greatest respect;
Boito (1836–1914)	New forms are distinguished from old forms, highlighting the modernity of the forms; Restoration Charter.
Cesare Brandi (1906–1988)	Safeguard value other than functional value.
Torres Balbás (1888–1960)	Respect the passage of time; understand the building as a historical document; preserve rather than restore; use recognizable materials to facilitate their subsequent identification; conduct a historical study of the building; Athens Charter.

Defined by the NBIMS-US (National BIM Standard-United States Project Committees), the BIM methodology consists of a digital representation of the physical and functional characteristics of a space system and a shared knowledge resource of information on a space system, which forms a trustworthy and reliable basis for decision making during the life cycle of a space, from its earliest conception to its demolition [13]. BIM is a collaborative system that is currently fully developed in the design and management of the industries involved in the architecture, engineering, and construction (AEC) sector. However, in the area of interventions in cultural and architectural heritage, there are very few studies dedicated to managing information models [14]. Although it is true that research has already been carried out to explore the value of BIM in the management of heritage and cultural landscapes [15], these are still scarce and continue to be a new topic in which there is still much to delve into, particularly in the case of industrial heritage. There are many arguments that corroborate the application of BIM to this type of case and to the creation of new buildings.

One of the major novelties offered by BIM is the inclusion of new dimensions for better control and management of the model. In this way, 4D refers to time, planning, and allowing a 3D model to be navigated as if it were a timeline, thus also enabling the control of the life cycle of the modeled building [16], while 5D refers to the economic aspects of the project (facilitated by the so-called parametric objects, Garagnani and Manferdini [17]), and 6D refers to sustainability [18–20].

The essence of BIM is collaborative, coordinated, and efficient work that consists of a virtual representation in a 3D model of an environment, a structure, or a building together with its intrinsic physical and functional characteristics. The model generated as a combination of an accurate geometric representation of the building in an integrated data environment is one of the main advantages of this methodology [21]. This advantage may be of special interest in the case of an intervention in industrial heritage real estate [22].

Implementing BIM is of global importance, not only because of the advantages it provides, but also because of its role in tender procedures in public projects. In Spain, the use of BIM is already mandatory in some public tenders [23]. Industrial heritage interventions fall within that obligation, which, together with the advantages of the use of BIM and heritage BIM (HBIM), can make it necessary and timely to teach this topic in engineering schools. The implementation of BIM in industrial engineering study plans can allow students to learn and acquire skills related to requirements of the real practice of engineering. In this sense, Caballero et al. [24] describes an implementation of the BIM methodology in project subjects for the degrees in mechanical, electronic and automatic engineering, electricity, industrial organization, industrial technologies, and chemistry in a school of industrial engineering based on a methodology of collaborative learning (CLM). Badak arranges the advantages of using CLM in three groups: social, psychological, and academic advantages. The following are a few of the advantages of CLM: It "establishes a positive atmosphere for modeling and practicing cooperation, promotes critical thinking skills, models appropriate student problem solving techniques, and is especially helpful in motivating students in specific curriculum and alternates student and teacher assessment technique" [25].

The present work is part of the BIM methodology implementation process in a school of engineering. The goal of this new step is to establish a line of work for the Final Project Degree (FPD) in engineering in a BIM context. Based on BIM as a project methodology and not as the adoption of a particular technology, during the development of an intervention project on a historic building, students have been integrated into a conceptually complete BIM process, participating in and undergoing the necessary stages for its completion, representing another step forward in the process of implementing BIM in schools of industrial engineering. This new phase of searching for common ground between the incorporation of the skills inherent to BIM and educational practices intends, in this particular case presented in this work, to establish sustainable intervention for an existing Spanish industrial heritage complex.

The European and national guidelines for preservation, including its reconversion proposal, have been taken into account. For the materialization of the experience, consisting of group and collaborative work, an interdisciplinary team of students from different undergraduate programs at a school of industrial engineering was formed; the students developed BIM models and other documentation to develop a complete project. The final work was constituted by the subprojects developed by each student. The true essence of BIM is its great collaborative power; therefore, without making use of this characteristic of the system, a conceptually integral project would not be achieved in any case, remaining in the partial stages of BIM in all cases.

With this intention, the main academic objective that guided us at the time of formalizing the present work was the establishment of a group work methodology that integrates engineering students in a conceptually complete process. For this, it is necessary to undergo all the stages needed to achieve such a project as well as to consolidate the concept of BIM as a collaborative project methodology and not as the adoption of a particular technology. In this environment, it is essential to organize information management in a context of collaboration between the actors involved and to develop a collaborative spirit and ethical behavior during debate and discussion.

The implementation of this work arises from the need, which is still being reaffirmed today, for the protection of industrial heritage as a witness to the historical, social, and economic past of territories. Many projects are currently at work in this context, not only for the conservation of these buildings, but also for their enhancement by converting them into new centers of social, cultural, and economic growth. The development of this project is framed within these lines, in which BIM is used in an academic environment, thus proving the concordance that exists between the preservation of an industrial heritage building and the use of BIM.

2. Materials and Methods

The academic experience described in this work consists of conducting a group and collaborative End of Course Project Proposal in a BIM environment. To complete the project proposal, an interdisciplinary team consisting of four students with different degrees in industrial engineering (2 students pursuing an electrical engineering degree, 1 student pursuing an industrial design engineering degree, and 1 student pursuing a mechanical engineering degree) and 4 teacher-tutors is formed. This team is divided into 4 groups, each composed of a student and his or her professor. Each of the students produces the BIM models and the other documentation required to develop a complete project (subproject). Together, all of the subprojects constitute the FPD for each student. The proposed project is completed by combining the 4 subprojects. The content of the proposed project consists of the intervention on an industrial complex belonging to the Spanish Industrial Heritage.

2.1. BIM Implementation in Final Projects in Engineering Degrees

The work presented in this article completes the implementation of BIM initiated in technical project subjects in an industrial engineering school. This implementation extends to the FPD of the engineering school so that students can acquire a BIM maturity level [26] greater than that obtained in the degree subjects in which BIM is already implemented.

FPD is a mandatory subject of the engineering degrees, and it is the last step to complete engineering degree studies. The FPD consists of an autonomous project developed by the student under the supervision of a tutor, where the student implements the skills of the degree and develops skills related to the student's specialization. The FPD has 12 credits and occurs in the last four months of the curriculum. The student's final grade in this course is given for the student's public presentation of his/her work to a court of 3 teachers. The court assesses the presentation and the defense of the results obtained, as well as the project documents produced by the student.

The implementation of BIM is developed within a collaborative methodology of active learning. The methodology begins with the presentation of the proposed project to the students, along with the collaborative methodology between teachers (FPD tutors) and students that will be used for the development of the project and the preparation of the BIM Execution Plan (BEP) (step 1). BEP specifies in detail the assignment of BIM uses, roles and responsibilities, deliverables, types of models, software versions, temporal planning through phases and milestones, planned collaborative work, etc. All the documentation generated in the BEP during this phase was allowed to be accessed by the entire team (students and tutors) in multi-platform cloud file hosting services.

In the face-to-face sessions, the different tasks to be performed by each student are presented, together with the necessary theoretical-practical explanations by the teachers (step 2). Through the student's autonomous learning and the queries that the student makes to his/her tutor, each student addresses his or her doubts and presents his or her proposal as a solution for the tasks assigned to them (step 3). The proposals of each student are discussed in a face-to-face session with the whole team (students and tutors) (step 4), giving way to the proposition of new activities when a consensus is reached on the proposed solutions (step 5). This process is repeated until all the tasks planned for the proposed project are completed (Figure 1).

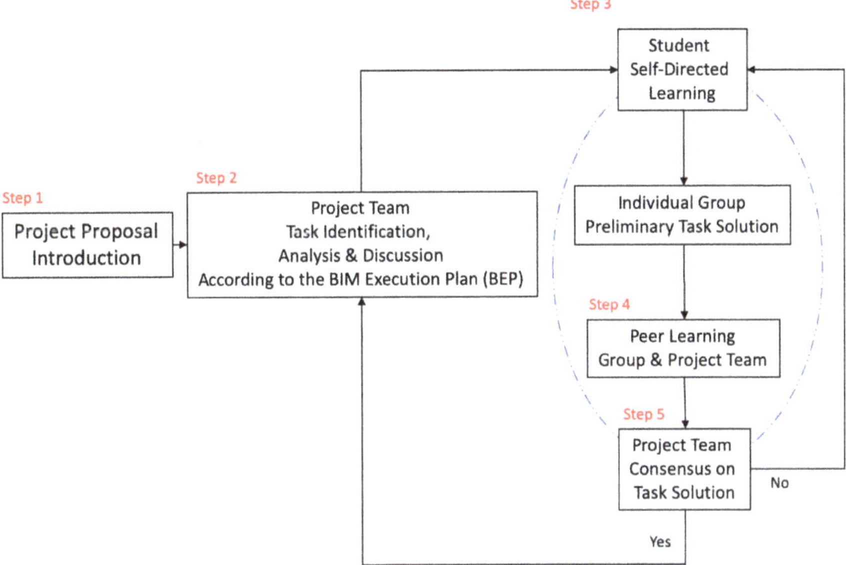

Figure 1. **Building information modeling** (BIM) implementation in final project degree subject: Overview of the active collaborative learning methodology applied.

The main tasks of the project are included in three phases: Recording the building's information (phase 1), modeling the existing building (phase 2), and designing the intervention as a proposal for the new use of the building (phase 3). These phases were developed in four steps (Figure 2).

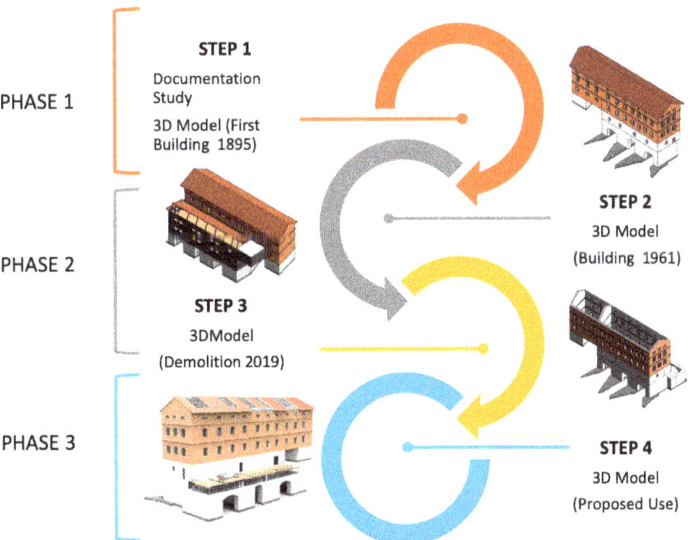

Figure 2. Phases and steps proposed in the real estate rehabilitation of an industrial heritage site using BIM for the case study presented.

For planning the new-use design, the location and analyses of the site, i.e., from studies on the topography, vegetation, and climatic conditions, were taken into account [20].

To carry out the project, the following BIM software was used: Autodesk Revit for 3D modeling and installation, Autodesk Robot Structural Analysis Professional for structural calculations, Archimedes (Cype) for the implementation of measurements and budgets, KeyShot for the renderings, and Navisworks to perform checks and detect collisions. Different scientific publications show the high utility of 3D models, renderings, and simulations to achieve the objectives outlined in heritage studies [27–29].

As for the collaboration procedures provided, which are the essence of BIM, due to the possibility of different locations of team members, a synchronized workflow was organized in real time in the cloud through the Dropbox platform. The coordinator of the work team was one of the tutor professors for the FPDs. Through Revit, the coordinator of the work team created a central file located in Dropbox, and each component created its local copy of the file, allowing simultaneous work in different locations through subprojects synchronized with the central model, following a rigorous systematic process.

FPD and its evaluation and feedback are an important part of engineering training [30]. Considering the score from the court responsible for the evaluation of FPD, the assessment by FPD tutors, and the point of view of the student in relation to the use of BIM and its acquisition of competencies, it has been possible to evaluate the results of this implementation in the FPD.

2.2. Project Proposal: Heritage Intervention as Final Projects in Engineering Degrees

The intervention focuses on the old flour mill "La Julita", which is located in Simancas Valladolid (Spain) and was built in 1865. It is an industrial complex of four buildings; the main building is located on the banks of the Pisuerga River, and the other three buildings function as warehouses. The main building, the factory body (Figure 3), is located perpendicular to the course of the river, and the three annexed warehouses are configured approximately parallel to the river. The present study focuses on the main building, as it is the most important and the largest in the group in terms of both surface and height. The building consists of four accessible floors plus a basement, which features canals and turbines to take advantage of the motive power of the water. To this factory body, years after its construction, without knowing the exact date, a two-story warehouse was added to its north face and in parallel, which we will study only in the context of its relationship with the main building. The mill was located in the main building, and most of the works were concentrated therein, especially those dedicated to transformation processes. Its surface, despite its trapezoidal geometry, resembles a 40×10 m rectangle. The auxiliary buildings, with two floors, are arranged at an angle greater than 90° with the main building. The location and configuration of the complex correspond to the need to take advantage of the motive power of the water in the river, which is necessary to carry out grain-grinding processes. The factory was active until 1960–1962 [31].

Following the BIM methodology proposed by Succar [32], it is necessary to generate an object-based model that allows correct collaboration on the network model. Focusing on heritage buildings, one of the great advantages offered by BIM is the possibility of working in phases. This system allows the reconstruction of the different stages of the building and the different interventions that may have occurred over time. In this way, it is possible to document not only a specific condition, but also the entire life of that work, and thus be able to access its characteristics at each point and at each moment; that is, it is possible to create a navigable timeline that narrates the tangible and intangible changes in the past and their parametric relationships over time, and projections for the future [15].

Figure 3. Photographs of the factory, 2019.

However, there are challenges when using BIM technology for industrial heritage, because BIM has to account for a number of particular characteristics and determine the differences compared to newly constructed buildings. To begin, if we compare a historic building with a recently constructed one, we can see that the list of its constituent elements is similar, i.e., both have doors, windows, enclosures, structure, etc., but the difference lies in the specific characteristics of each of these elements in one case and in the other. Therefore, the first issue is that historical objects/buildings have certain characteristics whose geometry and material are not usually represented in BIM libraries [13]. To address this issue, which is derived from the architectural complexity of many historic buildings, the HBIM approach has been developed. HBIM has been defined as the registration and modeling of existing buildings, generating a BIM geometry of point clouds [33]. Murphy defines HBIM as a new modeling system of historical structures that creates complete 2D and 3D models, which include details on the surface of objects in terms of their construction methods and material composition [34]. HBIM includes highly protected buildings that generally require broader intervention projects and careful management of their life cycle. Dore and Murphy [15] propose six HBIM elements: heritage documentation standards, technical data collection, 3D modeling concepts, "as-built" BIM, and procedural modeling. From HBIM, complete 3D models can be produced automatically, along with drawings with technical engineering language [13,34], documentation for the conservation of buildings (analysis of historical structures), and precise visualization [35].

The difference between existing common buildings and heritage buildings is that heritage projects consist of architectural, historical, and archeological documentation, which implies the technical reproduction of a context [1,36]. Project management of a historic building involves recording building information, modeling the existing building and its information, designing the intervention, developing construction works, and planning the conservation of the monument. All of these tasks are achieved through a multidisciplinary approach to the process [37].

Figure 4 shows the 5 criteria that the project team considered in the evaluation of proposals and the selection of solutions to the different tasks proposed in the project. Any proposal that does not meet criteria A (Fulfillment of the Eight Points of Modern Restoration) and E (Fulfillment of Law and Regulatory Requirements, Norms and Legislation) is rejected. The criteria of cost, profitability, and sustainability are evaluated using a Likert scale from 1 to 5, with higher scores given to solutions that perform better. Lower-cost alternatives are given a higher value. In the pooling with the entire project team, the alternative that meets criteria A and E and obtains the highest score by adding the scores of the other criteria is chosen.

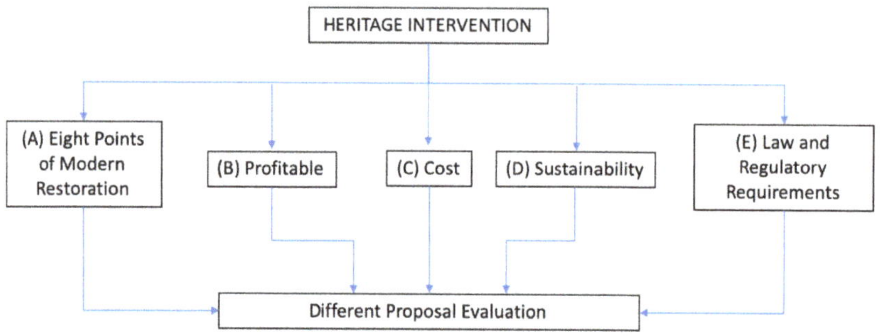

Figure 4. Proposal evaluation criteria.

3. Academic Feedback Results & Discussion

Students completed their FPD in approximately five months. All students' FPDs scored above average or better for their degrees. The results showed that students were able to demonstrate the use of BIM methodology and a positive impact on the acquisition of independent learning, problem-solving abilities, groupwork, decision-making, and communication skills. Macdonald also reports a positive impact, similar to that found in this experience, of the use of BIM and the acquisition of skills [38].

The applied collaborative methodology has been adapted from the methodology used in the second-year project subject taught at the engineering school where the experience described in this study has been carried out [39]. The results of this experience regarding the acquisition of competencies are in agreement with those obtained in the evaluation of this methodology by Blanco et al. [40].

Bower and Richards [41] describe peer review, mentoring, group-based tasks, and problem-based learning as options that involve some form of collaborative learning. These options have been used in this work, and our results are in agreement with Bower and Richards's results for groupwork skills.

The objective of the proposed study, using the collaborative methodology described in Section 2.1, is to develop a project. This point is common to methodologies such as project-based learning (PBL) [42,43] and cooperative-based learning [44]. The problem solving of some of the tasks proposed in the project, using the methodology of this study, is also a common element with methodologies such as cooperative problem-based learning (CPBL) [45].

The use of a collaborative methodology is not without barriers and limitations. Collaborative methodologies present, among others, the main barriers: construction, composition, and organization of working groups, design of activities to be carried out in groups, organizational guidelines, student learning pace, and evaluation of tasks. A systematic review of the scientific literature published in the last five years on the incorporation of collaborative methodology in different educational programs shows 11 different collaborative learning methodologies [46]. Revelo-Sanchez et al. [46] indicate that the application of a successful collaborative methodology consists of approaches and resources from various methodologies. In this line, as described, the methodology used presents elements of other methodologies that help reduce the barriers described. In this study, group size is very small, and students have previously taken courses in which they have used a collaborative methodology. These factors, together with the characteristics of the use of a BIM environment, make the limitations to the use of a collaborative methodology for the case proposed in this article smaller.

The use of the methodology described in this study assumes greater effort and dedication on behalf of the professor-tutor than the use of an approach with a non-collaborative methodology.

The FPD course aims to specialize the student in an engineering field. Different studies propose guiding FPD in a field of engineering specialization with good results [47–51]. Some authors propose the implementation of FPD with collaborating students [48,52,53], but not in the way in which it is proposed in this work within a BIM environment.

In previous courses, different students have performed FPDs aimed at a practical application of the BIM methodology. In all of these courses, the work was carried out by only one student, remained in the partial stages of BIM, and did not develop a complete model in any of the cases. The academic experience described in this study presents an approach that truly allows a group of students to collaborate in a BIM environment.

Students' perception was collected by professor tutors through interviews conducted with the students during the face-to-face delivery of their FPDs. Student's point of view in relation to their perceptions of their use of BIM and acquired skills indicates an average score of 4 (very satisfied) using a 5-point Likert scale on questions about to the independent learning, problem-solving abilities, groupwork, decision-making, communication skills, and assimilation of the BIM concept and its application. On the other hand, students' perceptions of the use of Revit indicate an average score of 3 (moderately satisfied). In relation to the acquisition of competencies, the use of BIM and its tools, these results follow a certain parallelism to those found by other researchers with students from other countries [54,55]

The incorporation of BIM in the project subjects in the engineering school in which the work described in this article has been developed is currently being evaluated. In relation to the point of view of the students in this experience, the results obtained in this study are in agreement with those already obtained in the study of the student's perception regarding the use of BIM and the assimilation of competences carried out in the advanced projects course of the final year of the degree in industrial technology engineering [56].

It is important to keep in mind that the academic proposal has been made in only one academic year. Currently, it has been proposed to conduct this academic experience in successive years to explore all the possibilities that this academic approach to the FPD subject can have.

4. Project Proposal: Heritage Intervention Results & Discussion

4.1. Phase 1. First Step: Original Condition

To reconstruct the original building, the work team searched for documentation in the Simancas Archive and at town hall, with practically no results. Basic graphic documentation was located in a publication on flour factories in the province of Valladolid; the document provided an initial orientation for the survey of the building [31].

As described by de la Red [31], stone, wood, brick, and ceramic tiles are, among others, the constituent materials of this factory, present in the foundation, interior structure, enclosures, and roof, respectively (Figure 4). Stone is used as the structural material, forming the foundation of the building, which has 1.20 m-thick walls. In the lower part of the building, three water channels run through arcades that are built with ashlars and continually exposed to erosion. In the interior structure, wood is used by means of insulated supports to support the weight transmitted by slabs of the same material. These supports are called "piedroits", which usually rest on a stone block and, at the top, are finished off with a wooden capital. However, wood acquires more importance in the architecture of the flour factory in the slabs for different floors. The beams are combined with filler ties, and planks are nailed on top, leaving the wooden floors visible. The roof, which is also made of this material, is defined as a row-to-row. Its use is rare because balance can be an issue; although it can be supported on wing walls, the central row or beam is generally supported by pairs of crossbeams. Wood is also used in the corbels of the roof eaves [31].

Aided by building documentation, the first step of the work methodology involved reconstructing the original condition of the building; notably, that there was no annex (Figure 5). The annexed warehouse was a later addition to the initial construction, which helps to differentiate the styles between the two parts of the complex. This addition is also a reflection of the probable changes that the then-established company underwent, assuming an increase in production that required an increase in warehouse surface.

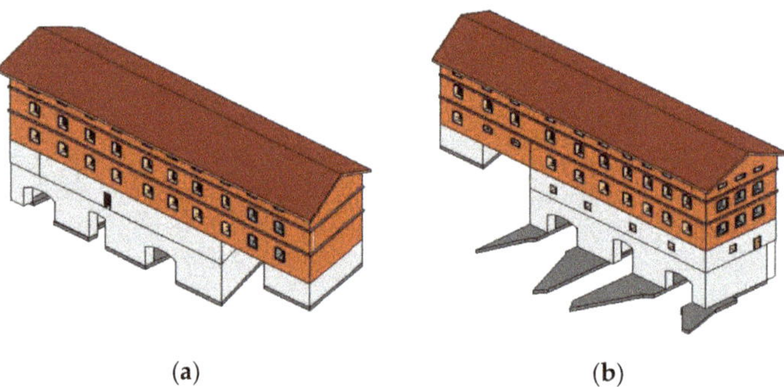

(a)   (b)

Figure 5. BIM model of the body of the factory: first phase, original condition. (**a**): Front view; (**b**) Rear view.

4.2. Phase 2. Second Step: Current Condition

Exhaustive fieldwork was carried out by performing topographic measurements to obtain real measurements of the building, and dimensions of the surrounding terrain, to perform BIM of the current condition of the industrial building.

Currently, more than half a century since its closure in 1961, the old flour mill "La Julita" is in a state of ruin and abandonment. Comparing the current and original conditions, many elements of its composition have been modified or destroyed by the passage of time. First, the terrain around the structure has undergone severe transformations that have practically covered the three canals, leaving only the one closest to the river open; however, water does not pass through them (Figure 6).

Figure 6. BIM model of the body of the factory: second phase, current condition. (**a**): Front view;
(**b**) Rear view.

Likewise, the first-floor windows are bricked, and the rest of the floors are empty openings
in which, in some cases, part of the old wooden frame persists; furthermore, many of the brick
Sardinia-style lintels have disappeared (Figure 3).

Compared with the brick, the exterior stone walls have not deteriorated much, due to the inherent
strength of the material itself. The interior wood structure is practically destroyed, the floor slabs
no longer exist, and only some wood structures remain connected to the exterior enclosures, interior
partitions, and the structure of the roof, which has nearly collapsed entirely. Due to the extreme
deterioration of the structure, there has been some attempt at reinforcement, due to the presence of
steel beams that protrude at the height of the second floor.

In this phase, the condition of the building was modeled as it currently exists (Figure 6). According
to the phase modeling, a warehouse addition was built after the initial construction, which allows a
comparison of the difference in styles between the two parts of the industrial complex. This addition is
probably a reflection of the probable changes experienced by the company, assuming that there was an
increase in production that required an increase in the storage area.

4.3. Phase 2. Third Step: Demolition Stage

Demolition is another phase recommended for conservation projects that aim to rehabilitate the
building for new-use development. This phase aims to show the condition of the building prior to
construction for new-use development. In this case, the model shows the maintenance of the exterior
structure, while the entire interior, with the exception of the transverse load wall, has been eliminated
due to its state of complete ruin (Figure 7).

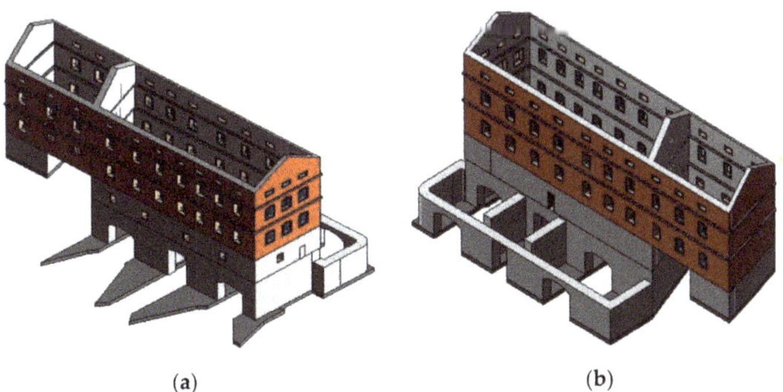

Figure 7. BIM model of the body of the factory: third phase, demolition stage. (**a**): Front view;
(**b**) Rear view.

4.4. Phase 3. Fourth Step: Good Practices

Finally, the last phase corresponded to the application of the rehabilitation proposal. Before beginning the process proposed in this work, it was necessary to establish an objective and contrasting approach to aid in decision making for this type of industrial heritage intervention.

As discussed in a previous section, one method of restoration with great international prestige and consolidation is the approach of the Italian Camillo Boito. In the Conference of Architects and Civil Engineers of Rome in 1883 [57], Boito presented his ideas on how the restoration of cultural assets should be carried out through the "Charter of Restoration", bearing the eight points of Modern Restoration, the predecessor to Scientific Restoration (Charter of Athens 1931). Its maxim was to "consolidate before repairing and repair rather than restore" to conserve, to the greatest extent possible, the peculiarity and originality of the good to be intervened.

The difference in styles between the old and the new, the difference in materials used in the work, and highlighting the authentic value of the building during the intervention are some of the eight points designed for all types of cultural property, with their own characteristics. Therefore, it is necessary to apply the eight points to the reality of this building and its peculiarities and possibilities.

This last phase of the intervention not only allowed for the comparison of different design proposals, but also verified decisions made and observed possible execution errors through BIM software used to simulate the real conditions of the building, and to identify possible conflicts in the project and the consequent construction. Regarding decoration modeling, the building of this practical case needs a 3D model that is less complex than other studies [58].

From the extensive study performed on the condition of the factory and its surroundings, as well as the circumstances that define and condition it, a series of guidelines were extracted that guided both the choice of the new use and the design proposal for its rehabilitation. Among the ideas that were established as objectives to be met, which we understand as best practices in the treatment of industrial heritage, we highlight as fundamental those that showcase the past history of the building and what occurred inside, as well as the context in which its activity developed; maintain the industrial character of the building and all its elements; ensure, in the new use and related rehabilitation process, the responsible use of resources contributing to the sustainable, social, economic, and environmental development of the environment; highlight the authenticity and originality of the building in contrast to the new use, which is defined by a different style and materials; and finally, protect the environment in which the building is located and its unique natural characteristics (Figure 8).

The process of a loss of function of the installations and productive infrastructures occurs rapidly for industrial buildings. When the factories cease to be productive, they also cease to be profitable; therefore, in most cases, adequate maintenance was not carried out for their conservation because no profit was obtained. When the industrial complexes that grew under the protection of industrial activities fell into disuse, they also fell into neglect. Factory buildings, as productive spaces that house obsolete productive activities, are disappearing and progressively falling into ruin. Therefore, it is a challenge to find new uses to which these goods can be associated [3].

Concerning the case studied here, Simancas is a municipality located approximately 14 km from Valladolid. It is a very important historic locale because of the General Archive of Simancas.

Another very valuable aspect of the municipality, from the environmental point of view, is the great ecological value of the Pisuerga River. The "La Julita" flour mill sits on the right bank of the Pisuerga River, very close to the confluence with the Duero River, which allows a reactivation of the use of the motive power of water for the energy supply to the building, contributing to sustainable development.

Consequently, to define a new use for the factory and therefore its new identity, the capacity of the factory to bring wealth to the community, both economically and culturally, was taken into account. This new use, as a priority, should change the negative significance that the community has regarding this building and its circumstances. To this end, a solution was sought within the environment, identifying a place with greater and better significance for the community and addressing a need not yet covered.

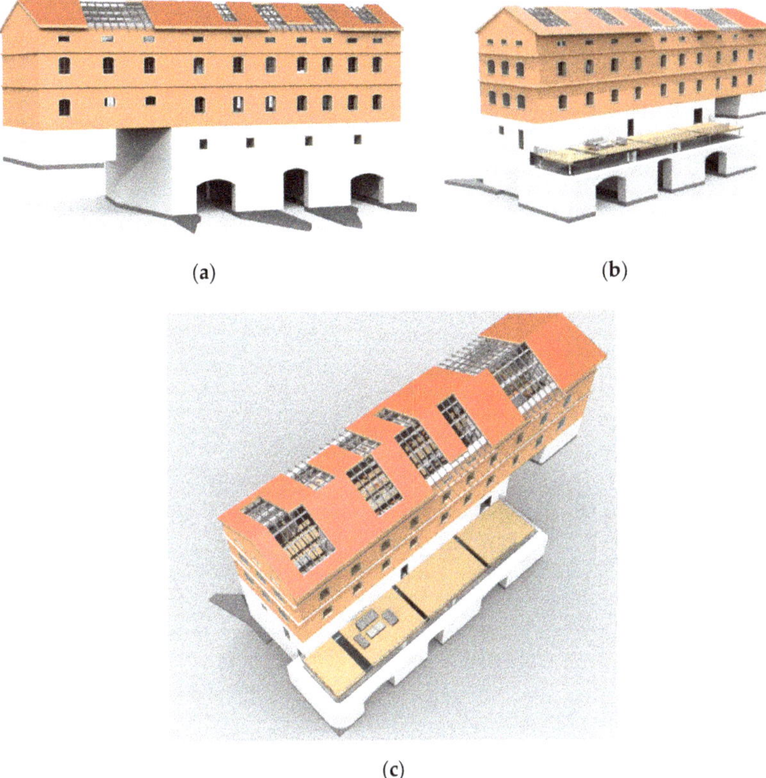

Figure 8. External renderings (BIM) of the proposed preservation. (**a**): Rear view (**b**) Front view; (c) Plant.

Finally, the new use was defined as a residential collaborative space whose objective is the promotion of research and collaborative work among professionals of all sciences, to promote the implementation of recovery projects and revaluation of industrial heritage buildings and goods. The residence plans included accommodations for researchers in connection with the General Archive of Simancas for consultation outside of its official facilities. Therefore, the objective was to establish a center for the creation of conservation and innovation initiatives in the field of industrial heritage protection. With this new use, the already known and integrated coworking concept can evolve into another context: Research.

Therefore, from the study performed on the condition of the factory and the environment in which it is located, as well as the circumstances that define and condition the factory, a series of results have been extracted as requirements that must guide the implementation of the new use as well as the design proposal for its rehabilitation:

- Spaces will be established to show the past history of the building and what occurred inside it, as well as the context in which it developed its activity. For this purpose, on the ground floor, a large part of the space will be dedicated to an exhibition of the past life of the building, with all its circumstances and characteristics, and the historical and economic context in which its activity developed. Likewise, space will be provided on each floor for sample activities that were carried out on that floor to assimilate each space with its function and original condition and contrast it with the new use.

- The industrial character of the building and all the elements that make it up will be maintained.
- Everything from the past will be preserved, respecting its authenticity and originality, that is, nothing of what remains will be modified, leaving imperfections that pose no risk.
- The industrial aesthetic will be preserved in the new-use development, but with the character of the current style, to achieve a pronounced contrast. What is added because of the needs of the new use will maintain the industrial character to the greatest extent possible: wide open spaces, concrete and defined uses of space, groups of simple and functional furniture, lighting from the roof, transit spaces, and evident circulation.
- With the new use, accessibility to cultural heritage in general will be promoted, as well as the protection of heritage assets of the complex in particular, both material and immaterial (processes, techniques, etc.), and their visualization for their knowledge. The use will be partially public, allowing visitors who do not want to use residency and coworking services access to all floors of the building and, thus, to the small sample activities on each floor; however, access will not be granted to the entire space on each floor, except the ground floor, which coincides with the largest exposition. Therefore, access to knowledge about the factory will also be guaranteed through accessibility devices needed to ensure full access.
- The responsible use of resources will be ensured in the new-use development and the linked rehabilitation process, contributing in turn to the sustainable, social, economic, and environmental development of the environment. The new-use development will incorporate, through current technology, the generation of energy through the motive power of water. Likewise, the interventions will focus mainly on conservation, limiting those of new construction to what is explicitly necessary. The new use will also fit perfectly with the possibilities of space: Large and ample spaces, good lighting, neutral colors, etc. It will also contribute to the economic growth of the community as a generator of jobs and wealth and, at the same time, will act as a new cultural space available to the entire society.
- Reconstruction, that is, the return to an earlier condition, will be considered as a final option. Everything that remains will be reinforced and conserved in its current condition, without reconstructing anything except in cases of fall risks, in which the necessary elements will be reinforced to mitigate the risk. Conservation of the original condition modified by the passage of time and neglect will be prioritized to help raise awareness about the importance of heritage protection, as it is not possible to recover what has been allowed to disappear.
- The environment in which the building is located and its peculiar natural characteristics will be protected. The environment will be maintained in its current condition, except for necessary cleaning and repair tasks. The first channel will reopen to harness energy from the river. In addition, within the exhibition established on the ground floor, there will be a section dedicated to explaining the natural environment and its importance with respect to the factory.
- The authenticity and originality of the building will be protected and highlighted in contrast to the new-use development, which will be defined with a different style and materials. Metal, glass and plastic will be used in an equally industrial style that contrasts with what already exists. New furniture will be perceived, to the greatest extent possible, as elements foreign to the place.
- The new use should promote a change in the significance that neighbors associate with the factory, from negative to positive. The new-use development will take advantage of the positive significance associated with the Simancas General Archive to revalue itself as a new cultural and research center of the municipality and contribute to economic development, which will also result in appreciation from the community.
- The new design will be aimed at making the most of the space, thus following the maxim *form follows function*. The design applied to the new-use development will be adapted to the existing conditions. Therefore, each plant will have a specific function, and the design will be adapted to take full advantage of the space and existing elements for the performance of the same, with the addition of new construction components if necessary.

Thus, the intervention focuses on the following aspects:

- Each floor has a specific function, similar to when it functioned as a flour factory;
- Open, wide, and luminous spaces, and transit spaces on the central axis;
- Materials in contrast to the existing ones: plastic, glass, and metal, as a priority, and treated and stained wood;
- Metal as structural material to maintain the industrial character;
- Zenithal lighting to accentuate the verticality of the space;
- Simple and functional furniture (Figure 9), and clean lines and contours; and
- Reestablish the use of water as the energy supply for the building, contributing to sustainable development.

Figure 9. Interior rendering (BIM) of the proposed new-use development.

One of the most important requirements to which all others are subject is that of, through design decisions, ensuring accessibility to all types of people and nondiscrimination of any group, as well as their safety within the premises. To this end, the pertinent regulations, stated in the Technical Building Code as applied to Architectural Restoration works, through the Support Document for the

Basic Document on Use and Accessibility Safety DB-SUA Code: DA DB-SUA/2 "Effective adaptation of accessibility conditions in existing buildings" and Basic Document SUA: Use and Accessibility Safety [59], were taken into account in this regard.

The design decisions that mostly affect the interior of the building are the requirements referring to the spaces and users' interaction with them. These are enunciated in the so-called conditions of accessibility, which in turn are divided into functional conditions (accessibility on the outside of the building, between floors of the building, and on the building floors) and accessible elements (accessible accommodations, reserved spaces, accessible bathrooms, fixed furniture, and mechanisms). Special attention was paid to fulfilling the specifications related to spaces and construction elements such as bathrooms, kitchens and dining rooms, public spaces, entrances, ramps, points of care, switches, and furniture.

This rehabilitation proposal brings together, on the one hand, the effective exercise of an innovative design that is totally appropriate to the space and respectful of the legacy of a historic building. On the other hand, it achieves the total protection and conservation of a space and its surroundings through a process of valuation, not only of its constructive characteristics and its potential as a container, but also of all the experiences that were developed within it and that currently make up an equally valuable and necessary heritage to be preserved and promoted for its full protection and promotion.

5. Conclusions

The BIM methodology adopted for the development of this project has helped and contributed to progress and achievement of satisfactory results, thus proving its effectiveness, even with certain shortcomings, in the industrial heritage environment. In the case that concerns us, through the use of the BIM methodology, students progressively acquired a high level of understanding of the technical aspects of a project. Through a holistic approach, which is part of the essence of BIM, the students understood the project and its object, and learned to value it as a global entity and not as the sum of its parts.

The academic results obtained by the students and the point of view of the students by participating in this teaching experience lead us to believe that the experience can be extrapolated, within the final degree project course, to other fields of specialization different from that used in this study. The proposed methodology could also be extrapolated to other undergraduate degrees.

The work conducted in this academic proposal represents the initial phase necessary for a proposed intervention, and can be used as a guide for future interventions on industrial heritage buildings with these characteristics.

With this design proposal for the rehabilitation of "La Julita" flour mill, all the design requirements and the marked specifications are fulfilled and encompassed in seven general and indispensable actions: conserve, value, teach, distinguish, take advantage of, protect, and promote.

As explained above, one of the main novelties offered by BIM that makes it very useful, with respect to other methodologies, for the preservation of historic buildings is the possibility of creating a timeline of the different phases that the building underwent. This process allowed us to develop models of the different conditions through which the factory has passed, allowing it to narrate in a much more complete and effective way the life of the building and facilitating its management in the present and in the future, including maintenance tasks.

Conducting rigorous historical background research prior to the intervention and the elaboration of a BIM model containing real content of the industrial complex provided the tools and means to carry out necessary "historical reading" before the preservation of the ruins. The "historical reading" not only provided an iconographic description of the complex and its physical and natural environment, but also may facilitate interpretive iconography, according to Panofsky [60]. Interpretive iconography allows for the recognition and interpretation of the symbolic values of an era and its industrial activity, which will allow us to understand and address the intervention in the most respectful way possible.

Author Contributions: Conceptualization, A.S., C.G.-G., P.Z., Z.S., and B.T.; investigation, A.S., C.G.-G., P.Z., Z.S., and B.T.; methodology, A.S., C.G.-G., P.Z., Z.S., and B.T.; supervision, A.S., C.G.-G., P.Z., and Z.S.; validation, A.S., C.G.-G., P.Z., and Z.S.; writing, original draft, A.S., C.G.-G., P.Z., and Z.S.; writing, review and editing, A.S., C.G.-G., P.Z., and Z.S.

Funding: This research received no external funding.

Acknowledgments: The authors thank all the students who developed the project with their tutors. In addition, authors would like to thank the anonymous reviewers for their comments, which have helped to improve the article.

References

1. De Naeyer, A.; Arroyo, S.; Blanco, J. *Krakow Charter: Principles for Conservation and Restoration of Built Heritage*; Bureau Krakow: Krakow, Poland, 2000.
2. Muñoz, A. Leopoldo torres balbás y la teoría de la conservación y la restauración del patrimonio. *Papeles Partal Rev. Restaur. Monum.* **2014**, *6*, 55–82.
3. Juan, C.G.I.L.; Ángel, S.P.M. *Aproximación y Propuesta de Análisis del Patrimonio Industrial Inmueble Español*; UNED: Madrid, Spain, 2016.
4. Sobrino, J. *Carta de Sevilla de Patrimonio Industrial 2018. Los Retos del Siglo XXI*; Fundación Pública Andaluza Centro de Estudios Andaluces, Consejería de la Presidencia, Administración Pública e Interior; Junta de Andalucía: Sevilla, Spain, 2019.
5. Benito del Pozo, P.; Calderón, B.; Ruiz-Valdepeñas, H.P. La gestión territorial del patrimonio industrial en Castilla y León (España): Fábricas y paisajes. *Investig. Geogr.* **2016**, *2016*, 136–154. [CrossRef]
6. Guited, F.G. *Guía Fabril e Industrial de España*; Librería Española: Madrid, Spain, 1862.
7. Gútiez, A.C. Ministerio de Educación Cultura y Deporte. In *Plan Nacional de Patrimonio Industrial*; Secretaría General Técnica, Subdirección General de Documentación y Publicaciones: Madrid, Spain, 2015.
8. Tagil, N. *Carta de Nizhny Tagil Sobre el Patrimonio Industrial*; The International Council of Monuments and Sites (ICOMOS) and The International Committee for the Conservation of the Industrial Heritage (TICCIH): Paris, France, 2003.
9. Brandi, C. *Teoría de la Restauración. Versión Española de María Angeles Toajas Roger*; Alianza Roma: Madrid, Spain, 1993.
10. Ruskin, J.; Estarico, L.; Burgos, C.D. *Las Siete Lámparas de la Arquitectura*; El Ateneo: Buenos Aires, Argentina, 1944.
11. Viollet-Le-Duc, E.E. *Entretiens sur L'architecture*; Pierre Mardaga: Bruxelles, Belgium, 1986.
12. Balbás, L.T. *En Torno a la Alhambra*; Al-Andalus: Madrid, Spain, 1960; Volume XXV, pp. 94–110.
13. Logothetis, S.; Delinasiou, A.; Stylianidis, E. Building information modelling for cultural heritage: A review. *ISPRS Ann. Photogramm. Remote Sens. Spat. Inf. Sci.* **2015**, *2*, 177–183. [CrossRef]
14. Nieto, J.E.; Moyano, J.J.; Rico, F.; Antón, D. Management of built heritage via HBIM project: A case of study of flooring and tiling. *Virtual Archaeol. Rev.* **2016**, *7*, 1–12. [CrossRef]
15. Fai, S.; Graham, K.; Duckworth, T.; Wood, N.; Attar, R. Building information modelling and heritage documentation. In *Proceedings of the 23rd International Symposium*; International Scientific Committee for Documentation of Cultural Heritage (CIPA): Prague, Czech Republic, 2011.
16. Maltese, S.; Tagliabue, L.C.; Cecconi, F.R.; Pasini, D.; Manfren, M.; Ciribini, A.L.C. Sustainability assessment through green BIM for environmental, social and economic efficiency. *Procedia Eng.* **2017**, *180*, 520–530. [CrossRef]
17. Garagnani, S.; Manferdini, A. Parametric accuracy: Building information modeling process applied to the cultural heritage preservation. *Int. Arch. Photogramm. Remote Sens Spat. Inf. Sci.* **2013**, *5*, 87–92. [CrossRef]
18. Jiménez-Roberto, Y.; Sebastián-Sarmiento, J.; Gómez-Cabrera, A.; Leal-Del, C.G. Analysis of the environmental sustainability of buildings using BIM (building information modeling) methodology. *Ing. Compet.* **2017**, *19*, 241–251. [CrossRef]
19. Fadeyi, M.O. The role of building information modeling (BIM) in delivering the sustainable building value. *Int. J. Sustain. Built Environ.* **2017**, *6*, 711–722. [CrossRef]
20. Bonenberg, W.; Wei, X. Green BIM in sustainable infrastructure. *Procedia Manuf.* **2015**, *3*, 1654–1659. [CrossRef]

21. Azhar, S.; Hein, M.; Sketo, B. Building information modeling (BIM): Benefits, risks and challenges. *BIM-Benefit [en línea]* **2008**, *18*, 11.

22. Julián, J.E.N.; Moyano, J.; Delgado, F.R.; Antón, D. La Necesidad de un Modelo de Información Aplicado al Patrimonio Arquitectónico. In *1er Congreso Nacional BIM-EUBIM 2013, Valencia, 24 y 25 de Mayo*; Universitat Politècnica de València: Valencia, Spain, 2013.

23. Iglesias, M.Á.G. Ley 9. Ley 9/2017, de 8 de Noviembre, de Contratos del Sector Público, por la que se Transponen al Ordenamiento Jurídico Español las Directivas del Parlamento Europeo y del Consejo 2014/23/UE y 2014/24/UE, de 26 de febrero de 2014. 2017.

24. Caballero, M.B.; Pérez, P.Z.; Fernández-Coppel, I.A.; Lite, A.S. Implementation of BIM in the subject technical industrial projects—Degree in industrial technologies engineering—University of Valladolid. In *Project Management and Engineering Research*; Muñoz, J.L.A., Blanco, J.L.Y., Capuz-Rizo, S.F., Eds.; Springer International Publishing: Champaign, IL, USA, 2017; Volume 201, pp. 247–260.

25. Rouyendegh, B.D.; Can, G.F. Selection of working area for industrial engineering students. *Procedia Soc. Behav. Sci.* **2012**, *31*, 15–19. [CrossRef]

26. Poljanšek, M. *Building Information Modelling (BIM) Standardization*; JRC Joint Research Centre: Ispra, Italy, 2017.

27. Rojas-Sola, J.I.; López-García, R. Engineering graphics and watermills: Ancient technology in Spain. *Renew. Energy* **2007**, *32*, 2019–2033. [CrossRef]

28. Sola, J.I.R.; Fuente, E. La esclusa de émbolo buzo de agustín de betancourt: Análisis de su construcción mediante ingeniería asistida por ordenador. *Inf. Constr.* **2019**, *71*, 286. [CrossRef]

29. Saygi, G.; Agugiaro, G.; Hamamcioglu-Turan, M. Behind the 3D scene: A gis approach for managing the chronological information of historic buildings. *Multimodal Technol. Interact.* **2018**, *2*, 26. [CrossRef]

30. Tien, D.T.K.; Lim, S.C. Assessment and Feedback in the Final-Year Engineering Project. In *Assessment for Learning within and beyond the Classroom*; Tang, S.F., Logonnathan, L., Eds.; Springer: Singapore, 2016; pp. 125–136.

31. De la Red, M.Á.C. *Las Fábricas de Harina en la Provincia de Valladolid*; Caja de Ahorros Provincial: Valladolid, Spain, 1990.

32. Succar, B. BIM ThinkSpace. Part of the BIMe Initiative (Bimexcelence.Org). Available online: http://www.bimthinkspace.com/ (accessed on 20 September 2019).

33. Dore, C.; Murphy, M. Current state of the art historic building information modelling. *Int. Arch. Photogramm. Remote Sens. Spat. Inf. Sci.* **2017**, *XLII-2/W5*, 185–192. [CrossRef]

34. Murphy, M.; McGovern, E.; Pavia, S. Historic building information modelling—Adding intelligence to laser and image based surveys of European classical architecture. *ISPRS J. Photogramm. Remote Sens.* **2013**, *76*, 89–102. [CrossRef]

35. Dore, C.; Murphy, M.; McCarthy, S.; Brechin, F.; Casidy, C.; Dirix, E. Structural simulations and conservation analysis-historic building information model (HBIM). *Int. Arch. Photogramm. Remote Sens. Spat. Inf. Sci.* **2015**, *40*, 351–357. [CrossRef]

36. Claver, J.; García-Domínguez, A.; Sevilla, L.; Sebastián, A.M. A multi-criteria cataloging of the immovable items of industrial heritage of Andalusia. *Appl. Sci.* **2019**, *9*, 275. [CrossRef]

37. Jordan-Palomar, I.; Tzortzopoulos, P.; García-Valldecabres, J.; Pellicer, E. Protocol to manage heritage-building interventions using heritage building information modelling (HBIM). *Sustainability* **2018**, *10*, 908. [CrossRef]

38. Macdonald, J.A. A Framework for Collaborative BIM Education Across the AEC Disciplines. In *37th Annual Conference of Australasian University Building Educators Association*; AUBEA: Sydney, Australia, 2012; pp. 223–230.

39. Caballero, M.B.; Lite, A.S.; Garcia, M.G. Sistema de Aprendizaje Para Proyectos de Ingeniería. In *19th International Congress on Project Management and Engineering*; CIDIP: Granada, Spain, 2015; pp. 2351–2362.

40. Blanco, M.; Gonzalez, C.; Sanchez-Lite, A.; Sebastian, M.A. A practical evaluation of a collaborative learning method for engineering project subjects. *IEEE Access* **2017**, *5*, 19363–19372. [CrossRef]

41. Bower, M.; Richards, D. Collaborative Learning: Some Possibilities and Limitations for Students and Teachers. In *Proceedings of the 23rd Annual Conference of the Australasian Society for Computers in Learning in Tertiary Education*; Markauskaite, L., Goodyear, P., Reimann, P., Eds.; Sydney University Press: Sydney, Australia, 2006; pp. 79–89.

42. Mills, J.E.; Treagust, D.F. Engineering education—Is problem-based or project-based learning the answer. *Australas. J. Eng. Educ.* **2003**, *3*, 2–16.

43. Guedes, G.T.A.; Bordin, A.S.; Mello, A.V.; Melo, A.M. PBL Integration into a Software Engineering Undergraduate Degree Program Curriculum: An Analysis of the Students' Perceptions. In *Proceedings of the 31st Brazilian Symposium on Software Engineering*; ACM: New York, NY, USA, 2017; pp. 308–317.

44. De los Ríos, I.; Cazorla, A.; Díaz-Puente, J.M.; Yagüe, J.L. Project–based learning in engineering higher education: Two decades of teaching competences in real environments. *Procedia Soc. Behav. Sci.* **2010**, *2*, 1368–1378. [CrossRef]

45. Yusof, K.M.; Hassan, S.A.H.S.; Jamaludin, M.Z.; Harun, N.F. Cooperative problem-based learning (CPBL): Framework for integrating cooperative learning and problem-based learning. *Procedia Soc. Behav. Sci.* **2012**, *56*, 223–232. [CrossRef]

46. Revelo-Sánchez, O.; Collazos-Ordóñez, C.A.; Jiménez-Toledo, J.A. El trabajo colaborativo como estrategia didáctica para la enseñanza/aprendizaje de la programación: Una revisión sistemática de literatura. *TecnoLógicas* **2018**, *21*, 115–134. [CrossRef]

47. Sampaio, A.Z. Building Information Modelling (BIM) taught in a Civil Engineer school. In Proceedings of the 2015 10th Iberian Conference on Information Systems and Technologies (CISTI), Aveiro, Portugal, 17–20 June 2015; pp. 1–6.

48. González, I.; Calderón, A. Development of final projects in engineering degrees around an industry 4.0-oriented flexible manufacturing system: Preliminary outcomes and some initial considerations. *Educ. Sci.* **2018**, *8*, 214. [CrossRef]

49. Díaz-Obregón, R.; Nuere, S.; D'Amato, R.; Islán, M. Strengthening the Interaction of Art and Science through Rubric-Based Evaluation Models: The Final Degree Project of the Dual Degree of Engineering in Industrial Design and Mechanical Engineering. In *New Trends in Educational Activity in the Field of Mechanism and Machine Theory*; García-Prada, J.C., Castejón, C., Eds.; Springer International Publishing: Champaign, IL, USA, 2019; pp. 141–148.

50. Pereira, D.; Neves, L. Students' final projects: An opportunity to link research and teaching. In *Geoscience Research and Education: Teaching at Universities*; Tong, V.C.H., Ed.; Springer: Dordrecht, The Netherlands, 2014; pp. 237–252.

51. Demaria, M.; Hodgson, Y.; Czech, D. Perceptions of transferable skills among biomedical science students in the final-year of their degree: What are the implications for graduate employability? *Int. J. Innov. Sci. Math. Educ.* **2018**, *26*, 11–24.

52. Bennett, D.; Male, S.A. A student-staff community of practice within an inter-university final-year project. In *Implementing Communities of Practice in Higher Education: Dreamers and Schemers*; McDonald, J., Cater-Steel, A., Eds.; Springer: Singapore, 2017; pp. 325–346.

53. Roberts, P.; Modi, S.; Roubert, F.; Simeonova, B.; Stefanidis, A. Information Systems Undergraduate Degree Project: Gaining a Better Understanding of the Final Year Project Module. In *Information Systems: Development, Applications, Education*; Wrycza, S., Ed.; Springer International Publishing: Champaign, IL, USA, 2015; pp. 145–170.

54. Shelbourn, M.; Macdonald, J.; McCuen, T.; Lee, S. Students' perceptions of BIM education in the higher education sector: A UK and US perspective. *Ind. High. Educ.* **2017**, *31*, 293–304. [CrossRef]

55. Jin, R.; Zou, P.X.; Li, B.; Piroozfar, P.; Painting, N. Comparisons of students' perceptions on BIM practice among Australia, China and UK. *Eng. Constr. Archit. Manag.* **2019**, *26*, 1899–1923. [CrossRef]

56. Sánchez, A.; Gonzalez-Gaya, C.; Zulueta, P.; Sampaio, Z. Introduction of building information modeling in industrial engineering education: Students' perception. *Appl. Sci.* **2019**, *9*, 3287. [CrossRef]

57. Rivera, J. Tres restauradores de la arquitectura, boito, giovannoni y Torres Balbás: Interrelaciones en la Europa de la primera mitad del siglo XX. *Conversaciones Rev. Conserv.* **2018**, *1*, 155–175.

58. Carnevali, L.; Lanfranchi, F.; Russo, M. Built information modeling for the 3D reconstruction of modern railway stations. *Heritage* **2019**, *2*, 2298–2310. [CrossRef]

59. España. *Aplicación del CTE (Código Técnico de la Edificación) a las Obras de Restauración Arquitectónica*; Ministerio de la vivienda—Consejo superior de los colegios de arquitectos de España: España, 2009.

60. Panofsky, E. *El Significado en las Artes Visuales*; Alianza Editorial: Madrid, Spain, 1980.

Article

Three-Dimensional (3D) Modeling of Cultural Heritage Site Using UAV Imagery: A Case Study of the Pagodas in Wat Maha That, Thailand

Supaporn Manajitprasert *, Nitin K. Tripathi and Sanit Arunplod

Remote Sensing and Geographic Information Systems Field of Study, School of Engineering and Technology, Asian Institute of Technology, P.O. Box 4, Klong Luang, Pathumthani 12120, Thailand
* Correspondence: smanajitprasert@gmail.com; Tel.: +66-61628-9532

Received: 1 August 2019; Accepted: 28 August 2019; Published: 4 September 2019

Featured Application: UAV-SfM was applied for generating 3D pagoda models in Thailand.

Abstract: As a novel innovative technology, unmanned aerial vehicles (UAVs) are increasingly being used in archaeological studies owing to their cost-effective, simple photogrammetric tool that can produce high-resolution scaled models. This study focuses on the three-dimensional (3D) modeling of the pagoda at Wat Maha That, an archaeological site in the Ayutthaya province of Thailand, which was declared a UNESCO World Heritage Site of notable cultural and historical significance in 1991. This paper presents the application of UAV imagery to generate an accurate 3D model using two pagodas at Wat Maha That as case studies: Chedi and Prang. The methodology described in the paper provides an effective, economical manner of semi-automatic mapping and contributes to the high-quality modeling of cultural heritage sites. The unmanned aerial vehicle structure-from-motion (UAV-SfM) method was used to generate a 3D Wat Mahathat pagoda model. Its accuracy was compared with a model obtained using terrestrial laser scanning and check points. The findings indicated that the 3D UAV-SfM pagoda model was sufficiently accurate to support pagoda conservation management in Thailand.

Keywords: UAV; terrestrial; photogrammetry; pagoda; heritage

---

1. Introduction

During the last decade, UAV-imaging has developed rapidly, enabling advancements in three-dimensional (3D) modeling for preservation, documentation, and management of cultural heritage sites [1,2]. Terrestrial surveying using terrestrial laser scanners is accepted for producing high-quality cultural heritage images, while the advent of robotic total stations has enabled researchers to gather large quantities of data more easily than ever before [3–6]. Unfortunately, the use of such geomatics-based approaches requires considerable expertise and a large surveying budget [2]. Alternatively, unmanned aerial vehicles (UAVs) can be used to easily and feasibly acquire 3D models. Furthermore, imaging using UAVs can be conducted economically and offers an ideal means for surveying complex archaeological sites [7–9]. There are different forms of UAVs, including fixed-wing [10], multi-rotor [11], and gyro. Furthermore, current platforms such as quadrotor [12] are more robust than their previous forms, due to their vertical take-off and landing capability, insensitivity to varying environments, high mobility and stability, and ease of operation [13,14]. The combination of computer vision and photogrammetry present in [15] has the following advantages: (1) Remote control systems, permitting the UAV to be perfectly positioned to collect images at varying heights and angles; (2) different sensors, ensuring that different image types can be used, including infrared, visible-spectrum, and thermal images captured from both calibrated and non-calibrated cameras, and (3) high-quality outcomes, enabling a researcher to control the reliability and accuracy of the results.

This study focuses on the low cost, portability, and completeness provided by the unmanned aerial vehicle structure-from-motion (UAV-SfM) method for cultural heritage site management. Two pagodas in Wat Maha That, Thailand (Chedi and Prang) are considered as a case study. The results demonstrate the accuracy of 3D models obtained using ground control points (GCPs), checkpoints (CPs), and terrestrial laser scanning (TLS). The study aims to apply the UAV-SfM method, which is cost-effective, convenient, and provides acceptable accuracy for building the 3D models compared with traditional techniques, such as TLS. The constructed 3D model could be employed in different applications in the future, such as tourism and generating the rapid mapping of cultural heritage in a short period, while ensuring acceptable quality.

2. Previous Work

Several previous studies have sought to use UAV technology for archaeological projects. Among these, Brutto [16] used the microdrone MD4-200 and the Sensefly Swinglet CAM UAV to photogrammetrically record the Temple of Isis. The equipment used in this case consisted of two UAV systems with different performance and characteristics fitted using a global station. Nadhirah and Khairul [17] proposed the use of UAV photogrammetry as a tool to capture and generate a 3D model in Negeri Sembilan, Malaysia.

Several other authors have used UAVs with certain geotechnological approaches; for example, Ebolese [18] used UAVs to produce a high-resolution 3D model of Lilybaeum, the ancient city of Marsala in Southern Italy. Chiabrando [19] studied an archaeological site in Hierapolis, Phrygia, Turkey, using UAV photogrammetry to collect aerial images of the site. Stek [20] used an inexpensive method involving drones to collect particularly high-quality spatial geoinformation, which could be employed to study the landscape archaeology at Le Pianelle in the Tappino Valley of Molise, Italy. Bolognesi [21] used a remotely piloted aircraft system for aerial photogrammetry at the Delizia del Verginese Castle of Italy. This was combined with a terrestrial laser scanner to generate a 3D archaeological model.

Research indicates that UAVs can be used for low-altitude imaging and remote sensing of spatial data [7,22–24]. UAVs are used for exploring cultural heritage sites because they are reliable and easy to use [25,26]. The latest developments in photogrammetry technology provide a simple, cost-effective manner to create a relatively accurate 3D model from 2D images [27,28]. These techniques provide a new set of tools for cultural heritage professionals to capture, store, process, share, and display images and annotate 3D models in the field. A review by Colomina and Molina [7] showed that the use of UAVs in the exploration of cultural heritage is increasing, owing to the ease of use and quality of processed measurements.

3. Materials and Methods

3.1. Cultural Heritage Sites: Wat Maha That, Thailand

The Wat Maha That site, including the site of the studied pagoda, is located at Ayutthaya, which was historically one of Thailand's capitals and is now a UNESCO World Heritage site. The city is located 85 km directly north of Bangkok. A historical park forms an important part of the city and is home to several culturally significant ancient temples. The historical park was granted UNESCO World Heritage status in 1991. This was justified under criteria III, which held that the site could be deemed an excellent witness to a period in which a true national form of Thai art was developed [29].

As shown in Figure 1, Wat Maha That is situated east of the Ayutthaya Island. It was a temple that held royal status and was considered the most sacred during the Ayutthaya period. The main Mahathat Chedi (bell-shaped pagoda) and Mahathat Prang (towering corn-cob pagoda) pagodas (Figure 2) contain relics from Buddha. At one point in history, these were temples where the Supreme Patriarch (head of the order of Buddhist monks in Thailand) resided.

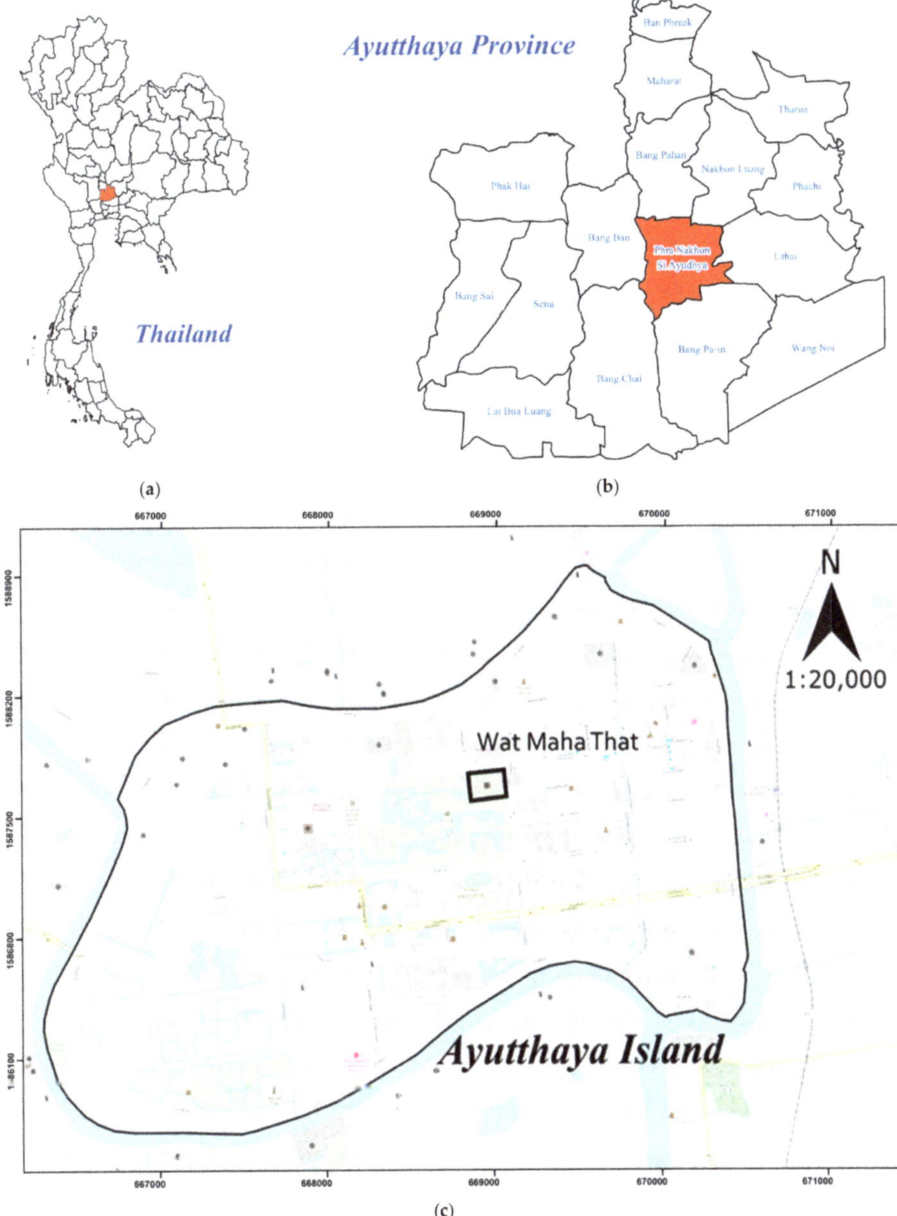

Figure 1. Aerial overview of the study area, Wat Maha That, Ayutthaya Island, Ayutthaya Province. (**a**) Map of Thailand. (**b**) Map of Ayutthaya Province. (**c**) Study area in Wat Maha That.

The Ayutthaya Chronicle [29] reported that the construction of the Mahathat Chedi commenced in 1374 during the reign of Phra Borom Rajathirat I and was completed under the reign of King Ramesuan. The foundation of the pagoda collapsed during the reign of King Song Tham, but was later reconstructed under King Prasat Thong's reign. Finally, the temple was destroyed and burned during the Burmese invasion in 1767 and has since been left in ruins.

(a)   (b)

Figure 2. Pagodas at Wat Maha That. (**a**) Prang structure. (**b**) Chedi structure.

3.2. Image Acquisition Using UAVs

In the study, the DJI Inspire 1 Pro UAV platform (Figure 3), a Zenmuse X5 digital camera fitted with a global navigation satellite system (GNSS) and an interchangeable lens that can be operated in the real-time kinematic (RTK) mode was used to take 417 images of the Wat Maha That.

Figure 3. DJI Inspire 1 Pro and Inspire 1 RAW platforms designed to assist in aerial imaging.

3.3. Flight Planning and Control

To collect high-quality data, it is necessary to carefully plan the UAV flight according to the technical limitations and attributes of the platform, its equipment, and sensors. To create the plan, it is first necessary to conduct visits to the site or gain access to detailed maps. A plan can then be created in stages:

1. A photogrammetric block is designed;
2. The flight path of the strips, along with their forward and side overlaps, are established;
3. The theoretical scale is determined using ground sample distance (GSD) calculations.

The photogrammetric block is determined by the size and location of the archaeological site and comprises take-off and landing points, along with the initial flight direction. From a theoretical perspective, the plan includes external orientation parameters (spatial and angular positions: X_0, Y_0, Z_0, ω, φ, and κ) and internal camera parameters (principal point and principal distance: X_0, Y_0, and c). The latter are set by the manufacturer of the camera. Once these parameters have been determined, it is possible to calculate the scale, flight altitude, side and forward image overlaps, number of strips, and number of images to be collected for each strip. If the platform does not feature gyro-stabilization, incorrect positioning will likely occur, necessitating the use of GNSS observations, i.e., GCPs (Figure 4). The final step is to ensure that the flight adheres to the original flight plan. This is accomplished using geometric controls, which ensure that the thresholds do not exceed certain criteria:

1. The vertical deviation of the image must be controlled by monitoring the angle between the optical axis and the nadir direction at the projection center.
2. Changes in direction and the drift effect must be controlled by monitoring the difference in the image coordinate system between the flight axis and the x-axis directions. In a planned flight, the theoretical value is $0°$ because the x-axis and the flight axis are in alignment.
3. The scale must be controlled. During digital photogrammetric flights, the definition of theoretical scale depends on the pixel size that is projected to the ground. Maintaining a constant GSD requires the ground to be a plane, but this is rarely the case in practice. The GSD is therefore dependent on the flight altitude and ground elevation, and can be calculated at any given point using a digital elevation model. The main aim is therefore to obtain an estimate for the GSD for each image and for each strip, which will eventually provide the mean GSD for the entire photogrammetric block upon completion of the flight.
4. The extent of the overlap must also be controlled. After calculating the scale and GSD, it is necessary to verify the forward and side overlaps that rise between images and between strips (Figure 5). This is essential, owing to the high degree of correlation between the altitude, GSD, and overlap values. The control of overlap depends on the verification of the side overlap between images and strips, and the forward overlap found between sequential images and strips.

Figure 4. Ground control points (GCPs) designed in a white and pink pattern.

As this particular platform type has no gyro-stabilizers and the GNSS is not highly accurate, it is necessary to use relatively wide limits to control the capture of images geometrically. Forward and side overlap limits are set at a minimum of 80% and 40%, respectively, while the error of the camera position is 3 m, and errors for the angular deviations are $0.2°$ (φ), $2°$ (φ), and $0.2°$ (κ). The greatest variation is 10% over the mean GSD size. Flight directional changes are restricted to a maximum of $2°$.

The flight was planned at an approximate altitude of 50 m, over an area of approximately 1 ha. The flight captured a total of 417 images in three west–east strips, with ten images per strip, a minimum forward overlap of 80%, and a minimum side overlap of 40% (Table 1).

Table 1. Specifications of unmanned aerial vehicle (UAV) flight planning.

Parameter	Value	Parameter	Value
Area of flight	1 ha	Baseline, b	32 m
Scale, S	1:50	Overlap, p	80%
GSD	20 mm	Sidelap, q	40%
Footprint	80×60 m	Spacing between strips, t	36 m
Flight height, H	50 m	Number of strips	3
Orientation of strips	West–East	Image per strip	10

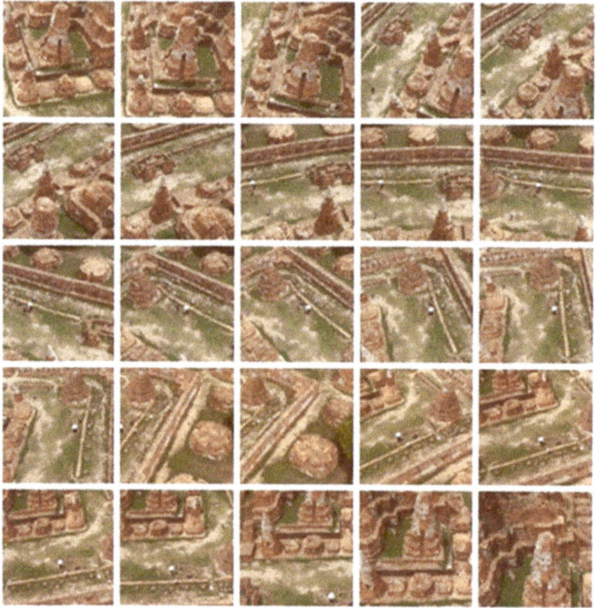

Figure 5. Overlap and sidelap images established from flight planning and control.

3.4. Reference Measurements

3.4.1. GCPs Measurements

GCPs were placed on natural features, such as the corner of the pagoda, and several on the interior of the interested area to assess the accuracy of the UAV-SfM method. Checkpoints (CPs) were used to check the pagoda accuracy in detail, and points were placed on the nature features of the pagoda façade.

3.4.2. Terrestrial Laser Scanning

The accuracy of a UAV-derived model was compared with that of the TLS data model, using a Riegl LMS-Z210 scanner recorded from multiple stations with high and medium resolution. TLS is a reference data model—its point cloud referred to known control points received from the RTK total station survey from benchmarks given by the Royal Thai Survey Department. The TLS device was set up at first at known XYZ coordinates, which were initiated from the total station. Finally, the result was applied as a reference for the UAV model. The accuracy of TLS is 3–5 mm. The 3D pagoda model was developed using the standard procedure implemented by CloudCompare software.

3.5. Image Processing

3.5.1. UAV Image Processing

A series of UAV images depicting an archaeological site were obtained, and these were then matched to specific points of interest. The hierarchical orientation of the images was the methodology adopted to produce an approximate orientation of the image, using the form of an arbitrary coordinate system through a computer vision technique. To automate image-based modeling and produce high-quality 3D point clouds, a structure-from-motion (SfM) algorithm was employed to process using Pix4D software. A total of 12 GCPs measured using a real-time kinematic GNSS were manually assigned to the corresponding locations on the textured model (Figure 6). The georeferenced data from the bundle adjustment between the recovered image blocks and the 3D model were optimized using the GCPs coordinates. We analyzed the quality of 3D models, using the coordinates that were measured independently of the ground truth point, and compared them with the photogrammetric coordinates. The 3D point clouds, textured meshes, and orthoimages were created to provide useful and accurate information for archaeological purposes (Figure 7).

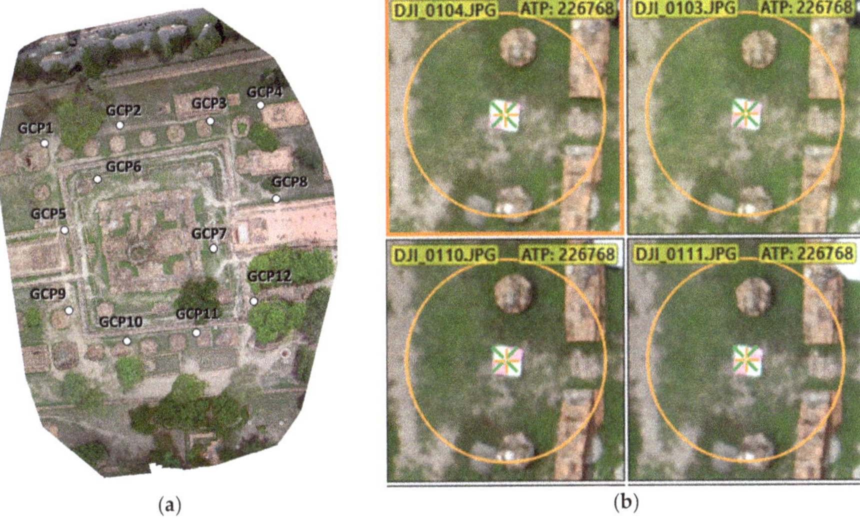

Figure 6. GCPs measured with a real-time kinematic global navigation satellite system (GNSS). (**a**) The locations of 12 GCPs. (**b**) GCPs marking.

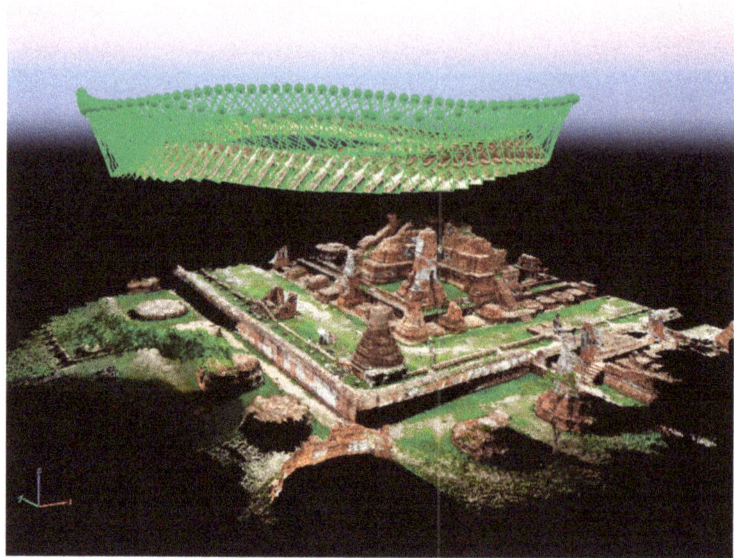

Figure 7. Camera position of the area of Wat Maha That, Ayutthaya flight.

3.5.2. 3D Pagoda Models Comparison

There were two datasets: One from TLS, which was the reference model, and the point cloud-produced nadir images from Pix4D. In order to compare the 3D pagoda model, point clouds were used to generate the 3D models using CloudCompare software. The 20 checking points scattered on the façade were mainly defined by using total stations to the physical structure of the pagoda. Obviously, the pagoda is a man-made structure, so some identical points can be measured and visually identified by the human eye; e.g., the pagoda's corners, windows, etc. However, the limitation of this study was not being allowed to place the checkpoint on the pagoda structure, due to the regulations and legislation governing historic structures with their conservative approaches, and the official suggestions were to declare it unsafe to climb the ruins of Wat Maha That.

4. Results

The initial UAV image processing of 417 images (0.016 m/pixel ground resolution at a flying height of 50 m) through the feature matching process, implemented in the SfM algorithm, produced a point cloud comprising 915,832 features over an area of 2020 m^2. The 3D UAV-SfM's model accuracy was evaluated by 12 GCPs, uniformly distributed over the study area using the following equations (Equations (1) to (3)):

$$\text{RMSE} = \sqrt{\frac{\sum_{i=1}^{n}\left(x_{\text{RTK},i} - x_{\text{computed},i}\right)^2}{n}} \tag{1}$$

$$\text{RMSE} = \sqrt{\frac{\sum_{i=1}^{n}\left(y_{\text{RTK},i} - y_{\text{computed},i}\right)^2}{n}} \tag{2}$$

$$\text{RMSE} = \sqrt{\frac{\sum_{i=1}^{n}\left(z_{\text{RTK},i} - z_{\text{computed},i}\right)^2}{n}} \tag{3}$$

where

RMSE	is the root–mean–square error
$x, y, z_{computed,i}$	is the point coordinates in the UAV images.
$x, y, z_{RTK,i}$	is the point coordinate measured from RTK.
n	is the number of GCPs

After bundle adjustment processing, the reported horizontal RMSE value was 0.028 m, and vertical RMSE was 0.230 m. The result is given in Table 2.

Table 2. Unmanned aerial vehicle structure-from-motion's (UAV-SfM) model accuracy assessment.

GCP	Field Survey Data			Diff		
	X (m)	Y (m)	Z (m)	dX (m)	dY (m)	Z (m)
1	669082.4191	1587840.2374	17.9260	0.021	−0.009	0.043
2	669064.9475	1587827.1337	19.0470	−0.012	0.012	−0.036
3	669010.3487	1587839.8734	18.5760	0.028	−0.029	−0.049
4	668978.6815	1587827.1337	18.1510	0.008	0.009	−0.086
5	668983.0494	1587790.0066	18.5300	0	−0.001	0.021
6	668996.8810	1587807.1142	18.2430	0	0.001	−0.081
7	669056.5757	1587778.3588	18.2350	0.047	0.003	0.026
8	669086.0590	1587798.7424	19.3790	−0.027	0	0.068
9	668984.1413	1587740.1397	20.1880	0.051	−0.040	−0.012
10	669018.7205	1587733.2239	19.8580	0.001	0	−0.057
11	669052.2078	1587737.2278	19.2190	0	0.001	0.021
12	669089.3349	1587739.7757	17.0110	0	0.001	−0.062
	RMSE H = sqrt (sum(dX)^2 + (sum(dY)^2/n)			0.028 m		
	RMSE V= sqrt (sum(dZ)^2/n)			0.052 m		

Furthermore, TLS and total station were performed to acquire reference data in this study for the independent coordinate checkpoints. To evaluate the overall quality positional accuracy of UAV-SfM results in a 3D pagoda model reconstruction, the 20 checkpoints were spread across the pagoda façade (bricks, window corner, etc.). Figures 8 and 9 show the distribution of these checkpoints on the pagoda façade.

(a)   (b)

Figure 8. Comparison of the terrestrial and UAVs point cloud computed on 20 checkpoints for the Prang structure. (**a**) Point clouds from terrestrial laser scanning (TLS). (**b**) Point clouds from UAV.

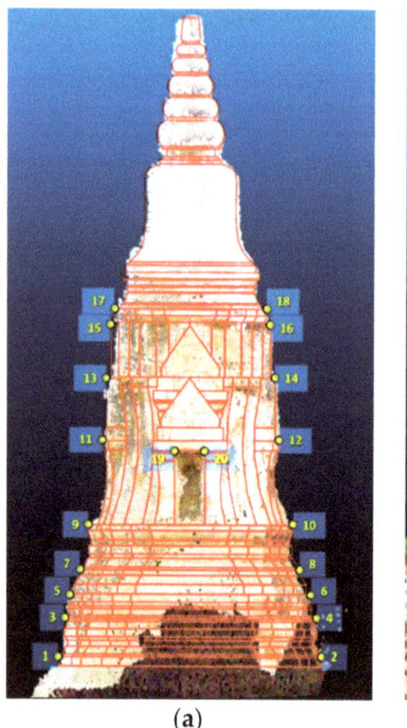

(a) (b)

Figure 9. Comparison of the terrestrial and UAV point clouds, computed on 20 checkpoints for Chedi structure. (**a**) Point clouds from TLS. (**b**) Point clouds from UAV.

4.1. First Case Study: Prang Structure

The models derived from the UAV and models obtained from TLS were compared with total station data to make more detailed assessments. The results listed in Table 3 indicate that the RMSE horizontal of the Prang structure for TLS-calculated data from 20 checkpoints of ground control is 0.068 m, and that of the UAV is 0.066 m. The RMSEs of the vertical Prang structure were 0.030 and 0.054 m for TLS and UAV, respectively.

Table 3. RMSE of Prang, computed on 20 checkpoints for the terrestrial and UAV point clouds.

	Terrestrial	UAV
RMSE H (m)	0.068	0.066
RMSE V (m)	0.030	0.054

4.2. Second Case Study: Chedi Structure

The results for the Chedi structure can be seen in Table 4. The RMSE horizontal for TLS was 0.069 m, and for UAV was 0.069 m. The RMSE verticals were 0.022 and 0.021 m for TLS and UAV, respectively.

Table 4. RMSE of Chedi, computed on 20 checkpoints for the terrestrial and UAV point cloud.

	Terrestrial	UAV
RMSE H (m)	0.069	0.069
RMSE V (m)	0.022	0.021

As shown in Figures 8 and 9, to evaluate the quality of positional accuracy on the pagoda façade, in this study, the selected checkpoints could be clearly seen in a field survey, such as natural features, the corner of the pagoda, and brick, and also clearly seen in UAV images and TLS data. The red line is a 2D representation obtained from the Fine Arts Department of Thailand's footprint to help assist in determining the checkpoint positions.

The reason why checkpoints are not selected at the top of the pagoda is that the edge or corner of the pagoda in the UAV and TLS point clouds cannot be clearly seen, as shown in Figures 10 and 11.

Figure 10. Comparison of the terrestrial and UAVs point cloud of the Prang structure: (**a**) Base of the Prang from TLS. (**b**) Top of the Prang from TLS. (**c**) Base of the Prang from UAV. (**d**) Top of the Prang from UAV.

Figure 11. Comparison of the terrestrial and UAVs point cloud of the Chedi structure: (**a**) Base of the Chedi from TLS. (**b**) Top of the Chedi from TLS. (**c**) Base of the Chedi from UAV. (**d**) Top of the Chedi from UAV.

5. Discussion

The UAV-SfM approach can generate orthoimages in a short time, with good quality and accuracy, and they are simple in nature, such that non-expert users can also use them. In this study, it was confirmed that SfM performances from UAV imageries could be implemented for modeling 3D cultural heritage. The UAV-SfM's application for pagoda patterns in Thailand provides archaeologists and local authorities with high-resolution geoinformation that can be used to support tools for rapid mapping of cultural heritage buildings.

The quality of GCPs for verifying and checking accuracy-based georeferencing depends on the number of GCPs and their distribution over the study area [30–32]. Our results showed that the calculated RMSE values of 12 GCPs were 0.028 m in the horizontal direction and 0.230 m in the vertical direction, which are acceptable results for georeferencing.

The accuracy comparison between the point clouds produced from UAV imagery and the terrestrial laser scanning data uses the mean error of 20 check points. The output of UAV on the pagoda compared with the TLS point clouds found that the calculated RMSE value of TLS and UAV point clouds were almost the same, owing to the comparable size of both the Prang and Chedi patterns. In addition, the distribution of checkpoints were only on the bottom base of the pagoda.

6. Conclusions

Developments in laser scanning technology were believed to have rendered the archaeological applications of photogrammetry obsolete; however, in reality, the techniques of digital image matching and image-based modeling are now considered a valid alternative to laser scanning. In this study, a highly effective and inexpensive tool is presented, which can create 3D models and orthoimages through the application of an image-based approach to modeling in the absence of other geotechnological equipment. This tool is flexible and easy to use, and employs proprietary software to generate images and models of complex scenes and objects, such as the archaeological site for the current demonstration at Wat Maha That.

This study evaluates the accuracy of the UAV-SfM method for pagoda exploration at Wat Mahathat, which helped determine the RMSE value as 0.028 and 0.052 m in horizontal and vertical directions, respectively. In addition, the results could be possibly applied in further analysis and for historical heritage management. The recording of these changes is particularly important for the architectural heritage of the Lord Buddha, where the relationship between personal architecture and the neighboring geographic environment has been maintained for hundreds of years.

Further work can investigate the integration of oblique images to enable the recording of historical objects in order to detect and record small details with high accuracy and completeness.

Author Contributions: Conceptualization, S.M. and N.K.T.; reviewed and edited the manuscript, S.M.; data curation, S.M. and S.A.; methodology, S.M.; software, S.M. and S.A.; validation, S.M.; formal analysis, S.M., N.K.T., and S.A.; writing—original draft preparation, S.M.; writing—review and editing, S.M. and N.K.T.; visualization, N.K.T.; supervision, N.K.T.

Funding: This research received no external funding.

Conflicts of Interest: The authors declare no conflict of interest.

References

1. Discamps, E.; Muth, X.; Gravina, B.; Lacrampe-Cuyaubère, F.; Chadelle, J.; Faivre, J.; Maureille, B. Photogrammetry as a tool for integrating archival data in archaeological fieldwork: Examples from the Middle Palaeolithic sites of Combe-Grenal, Le Moustier, and Regourdou. *J. Archaeol. Sci. Rep.* **2016**, *8*, 268–276. [CrossRef]

2. Xu, Z.H.; Wu, L.X.; Shen, Y.L.; Li, F.X.; Wang, Q.Z.; Wang, R. Tridimensional reconstruction applied to cultural heritage with the use of camera-Equipped UAV and terrestrial laser scanner. *J. Remote Sens.* **2014**, *6*, 10413–10414. [CrossRef]

3. Erenoglu, R.C.; Akcay, O.; Erenoglu, O. An UAS-Assisted multi-Sensor approach for 3D modeling and reconstruction of cultural heritage site. *J. Cult. Herit.* **2017**, *26*, 79–90. [CrossRef]

4. O'Driscoll, J. Landscape applications of photogrammetry using unmanned aerial vehicles. *J. Archaeol. Sci. Rep.* **2018**, *22*, 32–44. [CrossRef]

5. Themistocleous, K. Model reconstruction for 3-D vizualization of cultural heritage sites using open data from social media: The case study of Soli, Cyprus. *J. Archaeol. Sci. Rep.* **2017**, *14*, 774–781. [CrossRef]

6. Young, H.J.; Seonghyuk, H. Three-Dimensional Digital Documentation of Cultural Heritage Site Based on the Convergence of Terrestrial Laser Scanning and Unmanned Aerial Vehicle Photogrammetry. *ISPRS Int. J. Geo-Inf.* **2019**, *8*, 53. [CrossRef]

7. Colomina, I.; Molina, P. Unmanned aerial systems for photogrammetry and remote sensing: A review. *ISPRS J. Photogramm. Remote Sens.* **2014**, *92*, 79–97. [CrossRef]

8. Sayab, M.; Aerden, D.; Paananen, M.; Saarela, P. Virtual structural analysis of jokisivu open pit using 'structure-from-motion' unmanned aerial vehicles (UAV) photogrammetry: Implications for structurally-Controlled gold deposits in southwest Finland. *Remote Sens.* **2018**, *10*, 1296. [CrossRef]

9. Verhoeven, G. Taking computer vision aloft—Archaeological three-Dimensional reconstructions from aerial photographs with photoscan. *Archaeol. Prospect.* **2011**, *18*, 67–73. [CrossRef]

10. Aerial Data Systems, Fixed-Wing UAV. 2016. Available online: http://aerialdatasystems.com/fixed-wing (accessed on 17 January 2019).

11. Fallavollita, P.; Balsi, M.; Esposito, S.; Melis, M.G.; Milanese, M.; Luca, Z. UAS for Archaeology. New Perspectives on Aerial Documentation. *Int. Arch. Photogramm. Remote Sens. Spat. Inf. Sci.* **2013**, *40*, 4–6. [CrossRef]

12. Olson, K.G.; Rouse, L.M. A Beginner's Guide to Mesoscale Survey with Quadrotor-UAV Systems. *Adv. Archaeol. Pract.* **2018**, *6*, 357–371. [CrossRef]

13. Xia, D.; Cheng, L.; Yao, Y. A Robust Inner and Outer Loop Control Method for Trajectory Tracking of a Quadrotor. *Sensors* **2017**, *17*, 2147. [CrossRef] [PubMed]

14. Dong, J.; He, B. Novel Fuzzy PID-Type Iterative Learning Control for Quadrotor UAV. *Sensors* **2018**, *19*, 24. [CrossRef] [PubMed]

15. Fernández-Hernandez, J.; González-Aguilera, D.; Rodríguez-Gonzálvez, P.; Mancera-Taboada, J. Image-Based modelling from unmanned aerial vehicle (UAV) photogrammetry: An effective, low-Cost tool for archaeological applications. *Archaeometry* **2015**, *57*, 128–145. [CrossRef]

16. Lo Brutto, M.; Garraffa, A.; Meli, P. UAV Platforms for Cultural Heritages Survey: First Results. *ISPRS Ann. Photogramm. Remote Sens. Spat. Inf. Sci.* **2014**, *2*. [CrossRef]

17. Nadhirah, H.M.N.; Khairul, N.T. 3D Model Generation from UAV: Historical Mosque (Masjid Lama Nilai). *Int. Arch. Photogramm. Remote Sens. Spat. Inf. Sci.* **2017**, *42*, 251.

18. Ebolese, D.; Lo Brutto, M.; Dardanelli, G. UAV survey for the archaeological map of Lilybaeum (Marsala, Italy). *Int. Arch. Photogramm. Remote Sens. Spat. Inf. Sci.* **2019**, *42*. [CrossRef]

19. Chiabrando, F.; Nex, F.; Piatti, D.; Rinaudo, F. UAV and RPV systems for photogrammetric surveys in archaelogical areas: Two tests in the Piedmont region (Italy). *J. Arch. Sci.* **2011**, *38*, 697–710. [CrossRef]

20. Stek, T.D. Drones over Mediterranean landscapes. The potential of small UAV's (drones) for site detection and heritage management in archaeological survey projects: A case study from Le Pianelle in the Tappino Valley, Molise (Italy). *J. Cult. Herit.* **2016**, *22*, 1066–1071. [CrossRef]

21. Bolognesi, M.; Furini, A.; Russob, V.; Pellegrinelli, A.; Russo, P. Accuracy of Cultural Heritage 3D Models by RPAS and Terrestrial Photogrammetry. *ISPRS Int. Arch. Photogramm. Remote Sens. Spat. Inf. Sci.* **2014**, *40*, 113–119. [CrossRef]

22. Cho, G.; Hildebrand, A.; Claussen, J.; Cosyn, P.; Morris, S. Pilotless aerial vehicle systems: Size, scale and functions. *Coordinates* **2013**, *9*, 8–16.

23. Petrie, G. Commercial operation of lightweight UAVs for aerial imaging and mapping. *Geoinformatics* **2013**, *16*, 28–39.

24. Remondino, F.; Nocerino, E.; Toschi, I.; Menna, F. A critical review of automated photogrammetric processing of large datasets. *ISPRS Int. Arch. Photogramm. Remote Sens. Spat. Inf. Sci.* **2017**, *42*, 591–599. [CrossRef]

25. Themistocleous, K.; Agapiou, A.; Cuca, B.; Hadjimitsis, D.G. Unmanned Aerial Systems and Spectroscopy for Remote Sensing Application in Archaeology. *Int. Arch. Photogramm. Remote Sens. Spat. Inf. Sci.* **2015**. [CrossRef]

26. Remondino, F.; Barazzetti, L.; Nex, F.; Scaioni, M.; Sarazzi, D. UAV Photogrammetry for Mapping and 3D Modeling-Current Status and Future Perspectives. *Int. Arch. Photogramm. Remote Sens. Spat. Inf. Sci.* **2011**, *38*, 25–31. [CrossRef]

27. Ioannides, M.; Hadjiprocopis, A.; Doulamis, N.; Doulamis, A.; Protopapadakis, E.; Makantasis, K.; Santos, P.; Fellner, D.; Stork, A.; Balet, O.; et al. Online 4D Reconstruction using multi-Images available under open access. *ISPRS Ann. Photogramm. Remote Sens. Spat. Inf. Sci.* **2013**, *2*, 169–174. [CrossRef]

28. Koch, T.; Zhuo, X.; Reinartz, P.; Fraundorfer, F. A new paradigm for matching UAV and aerial images. *ISPRS Ann. Photogramm. Remote Sens. Spat. Inf. Sci.* **2016**, *3*, 83–90. [CrossRef]

29. Ayutthaya Historical Research. Available online: http://www.ayutthaya-history.com/Temples_Ruins_MahaThat.html (accessed on 16 January 2019).

30. Sanz-Ablanedo, E.; Chandler, J.; Rodríguez-Pérez, J.; Ordóñez, C.; Sanz-Ablanedo, E.; Chandler, J.H.; Rodríguez-Pérez, J.R.; Ordóñez, C. Accuracy of unmanned aerial vehicle (UAV) and SfM photogrammetry survey as a function of the number and location of ground control points used. *Remote Sens.* **2018**, *10*, 1606. [CrossRef]

31. Smith, M.W.; Carrivick, J.L.; Quincey, D.J. Structure from motion photogrammetry in physical geography. *Prog. Phys. Geogr.* **2016**, *40*, 247–275. [CrossRef]
32. James, M.R.; Robson, S.; Smith, M.W. 3-D uncertainty-Based topographic change detection with structure-From-Motion photogrammetry: Precision maps for ground control and directly georeferenced surveys. *Earth Surf. Process. Landf.* **2017**, *42*, 1769–1788. [CrossRef]

 *applied sciences*

Article

The Hay Inclined Plane in Coalbrookdale (Shropshire, England): Analysis through Computer-Aided Engineering

José Ignacio Rojas-Sola [1,*] and Eduardo De la Morena-De la Fuente [2]

[1] Department of Engineering Graphics, Design and Projects, University of Jaén, 23071 Jaén, Spain
[2] 'Engineering Graphics and Industrial Archaeology' Research Group, University of Jaén, 23071 Jaén, Spain
* Correspondence: jirojas@ujaen.es; Tel.: +34-953-212-452

Received: 10 June 2019; Accepted: 13 August 2019; Published: 16 August 2019

Abstract: This article analyzes the *'Hay inclined plane'* designed by the English engineer and entrepreneur William Reynolds and put into operation in 1792 to facilitate the transport of vessels between channels at different levels using an inclined plane. To this end, a study of computer-aided engineering (CAE) was carried out using the parametric software Autodesk Inventor Professional, consisting of a static analysis using the finite-element method (FEM) of the 3D model of the invention under real operating conditions. The results obtained after subjecting the mechanism to the two most unfavorable situations (blockage situation of the inertia flywheel and emergency braking situation) indicate that, with the exception of the braking bar, the rest of the assembly is perfectly designed and dimensioned. In particular, for the blockage situation, the point with the greatest stress is at the junction between the inertia flywheel and the axle to which it is attached, the maximum value of von Mises stress being at that point (186.9 MPa) lower than the elastic limit of the cast iron. Also, at this point the deformation is very low (0.13% of its length), as well as the maximum displacement that takes place in the inertia flywheel itself (22.98 mm), and the lowest safety factor has a value of 3.51 (located on the wooden shaft support), which indicates that the mechanism is clearly oversized. On the other hand, the emergency braking situation, which is technically impossible with a manual operation, indicates that the braking bar supports a maximum von Mises stress of 1025 MPa, above the elastic limit of the material, so it would break. However, other than that element, the rest of the elements have lower stresses, with a maximum value of 390.7 MPa, and with safety factors higher than 1.7, which indicates that the mechanism was well dimensioned.

Keywords: inclined plane; Coalbrookdale (Shropshire); computer-aided engineering; mechanical engineering; finite-element analysis; von Mises stresses; displacements; equivalent deformations; safety coefficient

---

1. Introduction

The work presented in this article was developed within a research project on the work of Agustín de Betancourt [1–3], analyzing his best-known inventions both from the point of view of engineering graphics [4–7] and from the point of view of mechanical engineering [8–12].

The article shows the static analysis carried out on the inclined plane for the transport of vessels that operated in Coalbrookdale (Shropshire, England) at the end of the 18th century. This historical invention, popularly known as *'The Hay inclined plane'*, and the work of the Englishman William Reynolds, made possible the transportation of boats between channels located at different levels.

This invention was the object of a very detailed study from the point of view of engineering graphics [13] that allowed us to obtain a reliable 3D model, from which the research presented in this article was carried out. There is no study from the point of view of mechanical engineering worldwide of this historical invention, a fact which underlines its novelty and scientific interest. The invention had a remarkable influence on the socioeconomic development of the region, as it was of considerable benefit for the coal mines and blacksmiths of the zone, since it allowed the commerce of their products through the river Severn, and the fluvial port Coalport on this river grew to be an important industrial hub [14].

The Shropshire canal was created in 1790 and closed to river traffic in 1944. This bi-directional inclined plane, which had a length of 350 yards (320 m) and allowed vessels to ascend and descend, was put into operation in 1792, and negotiated a difference of 210 feet (64 m).

Meanwhile, Agustín de Betancourt had dedicated much of his research to the study of navigation channels [1], and during his stay in England (1793–1796) he copied in detail the mechanism as it was built, drawing a color sheet of the inclined plane with much detail and a three-page handwritten account explaining the parts of the plane and its operation. These are the only documents of which there is evidence in reference to the inclined plane of Coalbrookdale [15]. The advantage of this historical invention was that it allowed a large drop to be negotiated without loss of water in the process of the ascent and descent of the boats.

Later, Agustín de Betancourt published his work 'Mémoire sur un nouveau système de navigation intérieure' written in Paris in 1807 [16], in which he proposes a channel navigation system for France very similar to the English navigation system: shallow channels and with an advanced system of locks that avoid loss of water in the ascent and descent of the boats. In this account, Betancourt applies his plunger lock to the inclined plane, and mentions the previous work of Robert Fulton [17], indicating that his inclined plane is designed according to Reynolds' procedure.

At present, only ruins remain of the inclined plane of Coalbrookdale. Some photographs [18] still show the inclined plane and the remains of the rails, although it is also possible to distinguish the upper cargo basin and the remains of the brick building that housed the steam engine.

The ultimate goal of this research is to perform a static analysis [19] of the *Hay inclined plane* using the finite element method (FEM) [20] under real operating conditions, in order to determine whether it was well dimensioned and functioned properly. Its scientific interest lies in the fact that from the point of view of industrial archaeology and the study of technical historical heritage, there is no existing study, worldwide, on this outstanding example of industrial historical heritage, which marked a historic milestone in the Industrial Revolution (1760–1840). This underscores the utility and originality of this research.

The remainder of the paper is structured as follows: Section 2 presents the materials and methods used in this investigation; then, Section 3 includes the main results such as von Mises stresses, displacements, deformations and safety coefficient; and Section 4 shows the main conclusions.

2. Materials and Methods

As mentioned above, the article has as its starting point the 3D model of the *Hay inclined plane* developed through our own methodology and published in a previous article [13]. Static analysis of this 3D model by means of FEM has been proposed.

Therefore, the sources used for the present study are the same as those used to obtain the 3D model. The Orotava Canary Foundation for the History of Science [21] developed a project on the engineer Agustín de Betancourt in which all his written and graphic work has been digitized [22]. In the file of the Spanish engineer [15], there is a written account and a detailed sheet of the existing inclined plane in Coalbrookdale that operated between the River Severn and the Shropshire Canal. The colored sheet, drawn by hand by Betancourt himself, describes most of the elements of William Reynolds' ingenuity, but cannot be considered as a scale plane itself, since without a written account it is impossible to understand the functioning of the mechanism for moving ascending and descending boats and, in addition, because some of the parts of the mechanism do not appear to be detailed in the plan, although they do in the written account.

2.1. Operation of the Hay Inclined Plane

To give the reader a complete idea of the analysis of the invention of William Reynolds, it is necessary to explain its operation. An exhaustive description of the mechanism is included in the article published on obtaining the 3D model [13], so a more summarized description will be given here.

Figure 1 shows an isometric view of the mechanism after its computer-aided design (CAD) modeling, and Figure 2 shows a plan of the ensemble with an indicative list of the different elements that form it. Figures 2–4 will serve to illustrate the operation of the mechanism.

Figure 1. Isometric view of the Hay inclined plane.

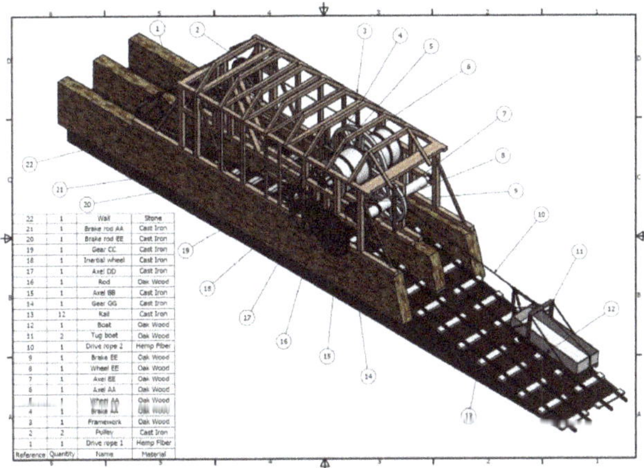

Reference	Quantity	Name	Material
22	1	Wall	Stone
21	1	Brake rod AA	Cast Iron
20	1	Brake rod EE	Cast Iron
19	1	Gear CC	Cast Iron
18	1	Inertial wheel	Cast Iron
17	1	Axel DD	Cast Iron
16	1	Rod	Oak Wood
15	1	Axel BB	Cast Iron
14	1	Gear GG	Cast Iron
13	12	Rail	Cast Iron
12	1	Boat	Oak Wood
11	2	Tug boat	Oak Wood
10	1	Drive rope 2	Hemp Fiber
9	1	Brake EE	Oak Wood
8	1	Wheel EE	Oak Wood
7	1	Axel EE	Oak Wood
6	1	Axel AA	Oak Wood
5	1	Wheel AA	Oak Wood
4	1	Brake AA	Oak Wood
3	1	Framework	Oak Wood
2	2	Pulley	Cast Iron
1	1	Drive rope 1	Hemp Fiber

Figure 2. Plan of the ensemble with an indicative list of the different elements.

2.1.1. Upward Movement

For the ascent of the boats (12), the tugboat (11) waited in the lower partially submerged part. The boat loaded with the material to be transported was placed on the tugboat, and they were tied together with a chain. After sending a signal to the operator of the upper station, a rope pulled the trailer, helping it to ascend on the rails (13).

The rope (10) that pulled the boat was picked up in the upper station on the AA axle drum (6) after going through some pulleys (2). As the axle rotated, the rope was coiled and the boat ascended. To move this axle, a steam engine was necessary, which moved a series of gears. An adjacent steam engine, of which only the chimney and the brick building where it was housed remain, operated an inertia flywheel (18). The wheel was solidly attached to the DD axle (17), which had two gears. The distal gear, which was the smaller gear, was coupled to the CC gear (19), which was the larger gear, which was solidly attached to axle AA, so that for each turn of the inertia flywheel the AA axle rotated 0.098 times in the opposite direction.

At the same time, the DD axle had a gear close to the inertia flywheel larger than the distal one. This gear meshed with an intermediate shaft, axle BB (15), which finally transmitted the movement to axle EE (7). For each turn that the inertia flywheel gave, the EE axle rotated, in the same direction, 0.14 times. A second rope pulling the tugboat that was in the upper channel was wound onto the EE axle drum.

To prevent the load from falling, or when it was necessary to move more slowly, two brakes could be used that braked the AA axle (4) or the EE axle (9) independently. For this, each axle had an inertia flywheel on which the brake was held, so that when pulling one of the braking levers (20) and (21), a collar embraced the wheels (8) and (5), respectively, and reduced the speed of the axles by friction.

2.1.2. Downward Movement

To descend the boats through the inclined plane, the process was similar. The path that had served to ascend the boat was automatically transformed into a descent route and vice versa. The engine had two hemispheres, where the ropes were wound onto the axles in the opposite direction, so while one of the axles was rolled, the other was unrolled with the same movement. When the rope was fully wound onto the drum, the end was taken on the other side of the drum, passed back through the pulley, and began to unroll.

The boat towed in the upper channel was pulled by the rope (1) attached to the EE axle to the upper station. In the upper station of the inclined plane, the rope was changed, and it was joined to the AA axle rope (10), which slowly began to unwind, allowing the plane to descend in a controlled manner. In cases of urgent necessity, both in the ascending and descending movement, the axle of the inertia flywheel DD (17) could be disconnected from the rest of the gears by actuating the shaft support bar (16).

Figure 3 shows the detail of the axles and gears that facilitated the movement of the ropes. It is interesting to note that each of the axles has an inertia flywheel on which is placed a friction collar that functions as a brake. It is also possible to appreciate the moving bar of the DD axle support that serves to leave the entire set of transmissions motionless if necessary.

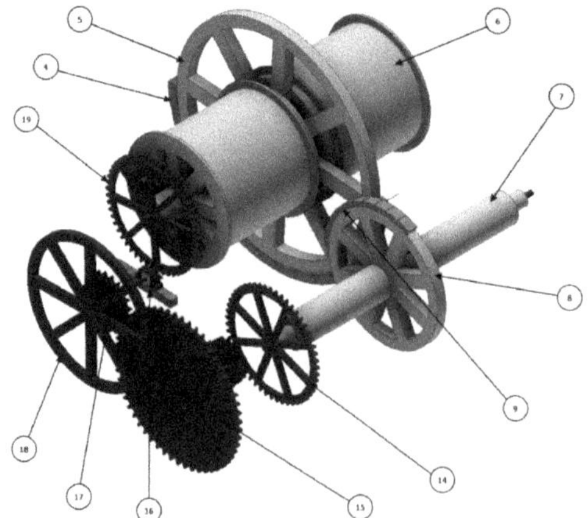

Figure 3. Detail of the transmission trees and gears of the engine.

Figure 4 helps describe the movement of axles and tugboats when they are in upward motion. In the case indicated, the tugboat on the right is ascending to the upper channel and the one on the left, at that moment empty, is directed to the lower channel.

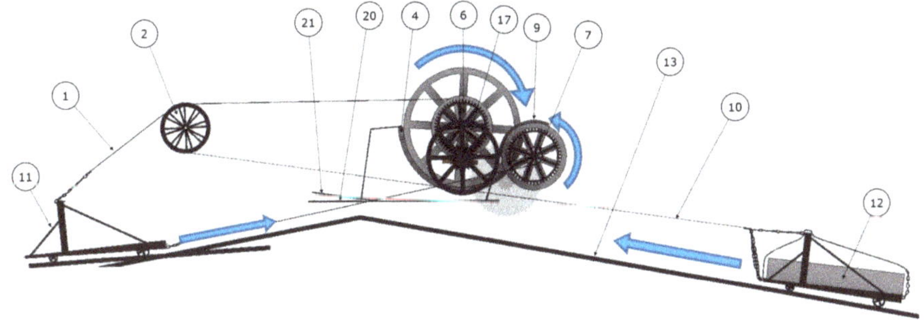

Figure 4. Sequence of movements in the ascent of the tugboats.

2.2. Computer-Aided Engineering (CAE)

Having defined and modeled the invention as designed by William Reynolds, it is now time to perform the static analysis with FEM, using Autodesk Inventor Professional software [23]. FEM is a numerical method used to solve differential equations by converting them into a linear system of equations. The results obtained by this method are only approximate, since a specific number of solutions are obtained for some points called "nodes", while in the rest of the points, the solution is interpolated, obtaining an approximate value. Likewise, the processes and parameters in a finite-element analysis (FEA) are independent of the software used.

This method pursues two objectives: the modal analysis that evaluates the natural frequency modes, including the movements of rigid bodies, and the static analysis of the invention, which evaluates its load conditions.

The numerical method used for static analysis generates a matrix with a large number of parameters, which can be categorized into three types: movement restrictions of the elements, the physical characteristics of the materials used, and the loads that affect it. On the other hand, the numerical model of the modal analysis is generated with fewer parameters, depending only on the restrictions and the physical characteristics of the elements, and therefore it is not conditioned by the parameters related to the loads.

Based on these data, the static analysis will obtain results on the von Mises stresses, and the equivalent deformations and displacements of all the elements of the model, while the eight lowest natural frequencies in which the elements of the model enter into resonance will be determined in the modal analysis.

Then the set of parameters that determine the analysis will be the simplification and definition of the model (pre-processing), assignment of materials, definition of boundary conditions and study of loads.

On the other hand, the "nodes" are not determined randomly. To obtain a result that represents the whole mechanism, a network of nodes called a "mesh" is established, which must be adapted to the geometry of each element that takes a similar form. This process is called "discretization", and each space delimited by the network is a "finite element". These elements have different shapes (surfaces, volumes or bars). To complete the analysis, it will be necessary to define the shape of the mesh (geometry and size), since the number of nodes depends, to a large extent, on the accuracy of the FEM. The more nodes the model has, the higher will be the precision, but also the calculation matrix and, therefore, the calculation time needed by the processor.

2.2.1. Pre-Processing

The first step before proceeding with the simulation of the static analysis is the pre-processing. The invention for ascending and descending boats presents a high number of elements, meaning that excessive computational time would be required to perform the calculations required by FEM. As will be indicated below, the software used estimates more than one million elements that must be analyzed, which would require a great deal of time on a personal computer with standard features.

Therefore, a simplification of the model is necessary (Figure 5) to facilitate the simulation. However, the simplification of the model should be performed with caution, as it should not interfere with the results of the static analysis. In addition, we must bear in mind that it is impossible to study the invention working in all possible circumstances, since these can all be very different. Taking these two premises into account, the static analysis of the inclined plane will be carried out in the two most unfavorable situations.

Figure 5. Simplified model of the Hay inclined plane for static analysis.

In the first place, the brick building, the rails and the plan have been dispensed with, focusing the study on seeing how both the engine and the support structure are affected. The roof, being a light structure exposed to the weather, has also been eliminated from the simulation, since it is not the support structure of the plane and it is not the object of the present study to analyze how it behaves in relation to the different types of external load (snow or wind).

Furthermore, it has also been decided to dispense with boats and tugboats. Each one of them presents a very high number of elements that would considerably multiply the calculations without affecting the results. Therefore, each of the tugboats has been replaced with a force that will later become the object of study.

The ropes that are associated with the two traction axles have also been dispensed with. The ropes of the time, made of hemp, are elements that had a special behavior. The software used is not prepared to correctly characterize its dynamic properties, which would force us to replace the tensions of the ropes with forces in the pulleys or by torques in the shafts.

Finally, it is necessary to examine the situations in which the engine is to be studied. On the one hand, the study of the invention was carried out with the boats fully loaded and ascending to the upper station. A priori, this case produces the maximum stress on all gears, pulleys and axles, so it is really convenient to study. However, the brakes, which are also important elements of the invention, would not come into play in this situation, so it was determined to study a second situation.

Therefore, a first situation would be to analyze the invention in the case of a blockage of the inertia flywheel, due to a disconnection with the steam engine, when the boat is fully loaded and descending. Another case, which would correspond to a second situation of emergency braking, would be the case of the tugboat with a loaded boat that had to be stopped using the brake that embraces the inertia flywheel of the AA axle.

2.2.2. Assignment of Materials

Having understood what aspects of the invention are to be studied, it is necessary to verify that all the elements of the assembly have their physical characteristics perfectly defined, without which the static analysis cannot be performed. However, the original materials of the engine are specified neither in the description of the invention nor in the written account, but it is not excessively complicated to determine with sufficient precision what they were. Thus, the metallic elements were deemed to be made of cast iron, probably forged in one of the ovens near the *Hay inclined plane* itself, the wooden elements would be of English oak, the rope hemp, and the constructed part of the installation of brick.

As mentioned in the previous section, neither the ropes nor the brick building form part of the static analysis, although the elements of the model have been perfectly defined. Certainly, the ropes are necessary in order to locate the direction of the tension they support on the axle, as will be discussed later, but they are excluded from the analysis for the reasons explained above.

Autodesk Inventor Professional has a very complete library of materials, where the physical properties (thermal, mechanical, elasticity and breakage) of each one are specified, although if necessary their properties can be defined or modified if the materials present some singularities.

In principle, the standard characteristics of the proposed materials have been respected. Thus, cast iron has the following values: Young's modulus (120,500 MPa), Poisson's ratio (0.30), density (7150 kg/m^3), and elastic limit (758 MPa). In contrast, oak has the following values: Young's modulus (9300 MPa), Poisson's ratio (0.0001), density (760 kg/m^3), and elastic limit (46.6 MPa).

Cast iron is an isotropic material, maintaining its physical properties in any direction, while oak is an orthotropic material, presenting a main working direction (direction of the grain), and different characteristics in the other directions. This main direction is the one that has been taken into account in the design of the support structure so that the beams and columns are modeled by choosing as the direction of the grain that in which the element has greater dimensions.

2.2.3. Boundary Conditions

The next step before starting the simulation of static analysis is to define the boundary conditions of the assembly elements. For this, the types of supports that exist in the structure are defined, as well as the manual contacts between certain parts.

Traditionally, the types of supports have been classified as embedded, articulated, mobile or rotating, but Autodesk Inventor Professional classifies them based on the degrees of freedom they possess. Thus, it defines fixed constraints on the surfaces of the elements that have no freedom of movement, sliding constraints on those for which one direction is impeded, and rotating constraints on the surfaces of the element that can only rotate around a certain axis.

In the model of the *Hay inclined plane*, it has only been necessary to define two types of constraints for the different surfaces: fixed and rotary. The fixed constraints are those that are applied mainly to the surfaces that are perfectly fixed to the brick building, such as the lower part of the wooden support (Figure 6a) and the support of the intermediary axle BB (Figure 6b).

On the other hand, the rotary constraints are applied to the surfaces that rotate freely, being fundamentally the points of contact between the axles and the supports on the structure. The transmission shafts have two supports, except for the DD axle, which has the inertia flywheel. This DD axle has only one point of support, because the inertia flywheel would be linked to the inertia flywheel of the steam engine that drives it through a belt or its own support structure. For this reason, the simulation will be carried out as if the surface of the inertia flywheel also served as a rotary support. The other elements that also have surfaces that rotate freely are the contact between the pulleys and the axle on which they roll, as well as the contacts between the first parts of each brake and the axle that supports it (Figure 7).

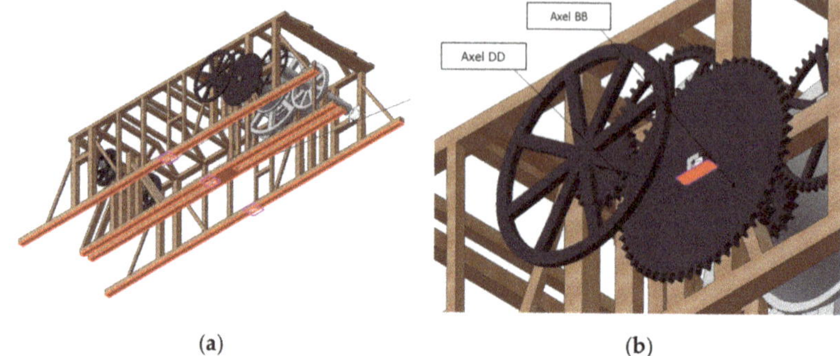

(**a**)   (**b**)

Figure 6. Fixed constraints applied: (**a**) support of the support structure (**b**) support of the BB shaft.

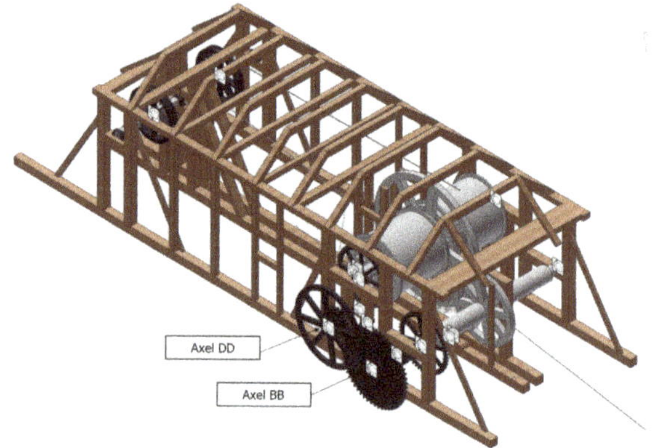

Figure 7. Rotary supports.

Regarding the manual contacts, it is especially important to check that the contact between the teeth of the different gears is correctly defined. Normally, when designing and positioning the gears, the edges of the gears do not come into contact, and therefore, the static analysis can be affected. Similarly, when studying the braking position, the contact between each of the brake parts and the inertia flywheel of the axle must also be correctly defined. Thus, in all cases, the defined manual contacts are of the blocked type (Figure 8), and the rest of the contacts between elements are recognized automatically by the software.

Figure 8. Manual contact defined between gears.

2.2.4. Forces Applied

The definition of the forces that affect the inclined plane is perhaps the most delicate step in the preparation of the simulation, since the various parts that compose it are affected differently by different forces.

Previously, the two situations in which the mechanism will be simulated were presented: a blockage situation of the inertia flywheel, and an emergency braking situation. Both present common forces, but also different forces that will be studied in these situations.

Among the forces common to both situations are the force of gravity and the tensions applied to the pulley in the direction of the rope.

Regarding gravity, the software defines it by specifying the magnitude and direction, and for it to be applied at the center of gravity of the engine, it is first delocalized by defining a generic vector of 9810 mm/s^2 in the direction of the Z axis in the negative direction (Figure 9).

Figure 9. Gravitational force applied at the center of gravity of the engine.

On the other hand, there is a tension applied to the pulley through the rope, and this is due to the weight of the tugboat along with the boat. The rope embraces the pulley on the outside, and therefore, it would not be necessary to characterize this tension if the behavior of the rope was simulated. However, as explained above, the dynamic behavior of the rope prevents it from being included in the simulation, raising two questions: on the one hand, how can the tension of the rope be simulated on the outer area of the pulley?, and on the other, could the load be replaced with a torque on the pulley shaft? The answer to the first question is affirmative, but not for the second.

To be able to apply the tension on the pulley in the most realistic way, a piece of rope was taken and joined together with the pulley. On the first contact point of the pulley with the lower and upper rope, two working planes were defined in a manner transverse to the main direction of the rope. On these planes, a load equivalent to the previously calculated tension was applied (Figure 10). In addition, for each of the tensions to have the proper direction, a larger piece of rope was used, although once the loads had been defined, it was excluded from the simulation.

The two tensions have the same value, because, as will be explained later, the AA axle exerts a moment of inertia that cancels out the load of the tugboat, so the situation is static. Due to this, the value of the tension is equal to the weight of the tugboat with the boat.

According to the data provided by Betancourt, the inclined plane is 210 feet high on a plane of 880 feet in length, which translates to an angle of inclination (α) of 13.806°. Furthermore, it is considered that the mass of the boat, using the materials of the time, is $M_1 = 2951.453$ kg (including the trailer and the chains that rig it), and in addition it is assumed that the boat is fully loaded with a mass $M_2 = 5000$ kg. Therefore, the value of the tension on the rope will be:

$$T_{rope} = (M_1 + M_2) \times g \times \sin\alpha = 7951.453 \text{ kg} \times 9.81 \text{ m/s}^2 \times \sin(13.806) = 18{,}614.44 \text{ kN} \qquad (1)$$

Figure 10. Forces applied on the pulley in the direction of the rope.

Forces at Play in the Situation of Inertia Flywheel Blockage

The calculation of the torques that cause the shafts to overcome the tension of the rope and thus to be able to raise the load on the plane is decisive. In addition, the torques of the AA and EE shafts do not have the same magnitude, since the slopes of the planes are different, although they are loaded equally. Because of this, they will be calculated independently.

The torque on the AA axle (Figure 11a) has to do with the previously calculated tension. Its calculation is very simple, since the value of the radius of the shaft drum is known, r = 1.185 m.

$$M_{AA} = T_{rope} \times r = 18{,}614.44 \text{ N} \times 1.185 \text{ m} = 22{,}058.11 \text{ Nm} \tag{2}$$

On the other hand, the torque along the EE axle (Figure 11b) has to do with the tension supported by the tugboat rope that ascends from the upper channel. It can be assumed, in the worst case of tension, that the tugboat is also fully loaded and, therefore, that the mass is similar to the aforementioned $M_1 + M_2$. However, the inclination of this plane is lower than the previous one, with a value of $\alpha = 10.470°$. With these data, the tension of the EE axle rope will be:

$$T_{rope\ EE} = (M_1 + M_2) \times g \times \sin \alpha = 7951.453 \text{ kg} \times 9.81 \text{ m/s}^2 \times \sin (10.470) = 14{,}174.89 \text{ kN} \tag{3}$$

Similarly, the radius of the EE axle is r = 0.3386 m, so the torque will be:

$$M_{EE} = T_{rope\ EE} \times r = 14{,}174 \text{ N} \times 0.\,3386 \text{ m} = 4799.62 \text{ Nm} \tag{4}$$

Finally, the inertia flywheel has to work as if it were blocked by forcing the gears in order to determine its resistance. Therefore, but only in this case, it is necessary to include a new fixed constraint to the external surface of the inertia flywheel (Figure 12).

(**a**)   (**b**)

Figure 11. (**a**) Torques applied on the AA shaft and (**b**) on the EE shaft.

Figure 12. Fixed constraint for the inertia flywheel of the DD shaft in the blockage situation.

Forces at Play in the Situation of Emergency Braking

The situation of emergency braking, in the event that the inertia flywheel of the steam engine were to break, for example, is an extreme situation that can occur, and that should be assessed for the correct dimensioning of the engine. In this case, it would only be necessary to stop the loaded tugboat that travels along the AA axle, since stopping the tugboat on the EE axle is not so important, as the distance it travels is short, and it would be immediately braked by the water of the upper channel with no danger of losing the load.

For this reason, only the torque on the AA axle, the tensions in the pulley, gravity and, finally, the brake force on the AA axle will be considered. This force has to be equivalent to the torque applied in order to be able to stop the load. Finally, in the case studied, the inertia flywheel would be released from the previously imposed fixed constraint and be able to rotate freely.

To characterize the AA axle brake, it would be necessary to study in detail the properties of the wooden brake that comes into contact with the inertia flywheel, specifically the friction coefficient of the wood. However, this coefficient changes due to a multitude of factors, such as humidity, possible surface scratching, or wear, among other causes, although an approximate value of the static friction coefficient of $\mu = 0.4$ can be given. This value, excessively low, is of great help in calculating the force that must be exerted on the brake bar AA to stop the tugboat in the case of breakage of the transmission mechanism.

The brake consists of 16 pieces with an approximate surface area of 0.111 m^2 each, which means a total braking area of 1.775 m^2. In addition, it is known that the force that must be exerted on the perimeter of the drum must be able to cancel out the torque applied on the AA axle with a value of M_{AA} = 22,058.11 Nm, according to Equation (2), but on his own inertia flywheel, which has a radius of value r = 2.511 m, which is greater than the radius of the drum.

With these data, it is now possible to calculate the pressure that the brake must exert on the inertia flywheel.

$$P = (F/S)/\mu = (M_{AA}/(r \times S))/\mu = 22{,}058.11 \text{ Nm}/(2.511 \text{ m} \times 1.775 \text{ m}^2 \times 0.4) = 12{,}372.66 \text{ N/m}^2 \quad (5)$$

Thus, the pressure that must be exerted is finally due to the force applied to the brake bar, and this is obtained by multiplying the calculated pressure P by the braking surface S.

$$F = P \times S = 12{,}372.66 \text{ N/m}^2 \times 1.775 \text{ m}^2 = 21{,}961.47 \text{ N} \quad (6)$$

As can easily be verified, this force (Figure 13) is equivalent to lifting something of more than 2 tons, so the loaded tugboat could not be manually stopped with the brake alone. This makes sense, since the trailer has a mass of around 8 tons. Thus, even if the coefficients of friction were higher and the force about half, it would still be excessive to brake manually. As has been demonstrated, although William Reynolds devised a complex system of levers and articulations to decrease the force necessary to stop a tugboat that is falling freely, the force that must be exerted continues to exceed the capacity of a single human operator.

Figure 13. Force exerted on the braking bar in the situation of emergency braking.

In any case, in this research, the behavior of the engine will be analyzed in a hypothetical case in which such a force could be exerted on the braking bar, although this situation would be highly unlikely.

2.2.5. Meshing

The last step before beginning the static analysis consists of the geometric discretization or meshing of the engine. Most software that performs an FEA uses tetrahedral elements, type tetra 10 (4 physical points and 10 nodes). If the simulation is more complex, for example, in dynamic analysis or in static analysis of complex geometries, some software uses hexahedral volumes, such as hexa 8, hexa 20 or hexa 27 for meshing. The use of the hexahedron allows the mesh to be formed by elements of greater volume, and therefore it needs a smaller number of elements to generate all of the geometry of the model. The choice of hexahedral meshes decreases the computational requirements and saves time, but results in less accurate results, since the total number of nodes is reduced and the mesh is less suited to the real geometry. In an analysis such as the present study, there is no problem in using tetra 10, since the total number of elements to analyze is not excessive.

Therefore, in FEA analysis, the elements used were tetrahedrons with 4 nodes and first order integration (constant interpolation of the stress and strain). The elements are formulated in a 3D scheme with three degrees of freedom per node (translational in X, Y and Z directions). However, a sensitivity analysis of the size of the element has not been carried out, since the structure is oversized.

On the other hand, the specific software used in the analysis of thermal parameters or lighting uses bar elements or shell elements. These elements of lower order than those used in the present study could also be used for the static or modal analysis of a mechanism, but it would require a much larger number of nodes to simulate its geometry, and therefore for a computer with a standard processor, these meshes would need an excessive time in their calculations. Regarding the shell elements, although they are capable of capturing the stress gradients in a more efficient way, they have not been used in this study, since the mechanism is oversized, and because it would provide an accuracy that would not change the conclusions in a significant way.

Autodesk Inventor Professional can automatically generate a mesh based on the size of the element to be discretized. By default, it generates tetrahedra whose average size is 10% of the length of the element, a minimum size of the tetrahedron of 20% of the average size, a maximum variation between tetrahedra of 1.5 and a maximum angle of rotation of 60°.

However, it is advisable to pay more attention to the places where the geometry is complex, since in these parts, the mesh usually gives results that differ greatly from the actual geometry of the element. For these singular places, the software offers the possibility of carrying out a manual meshing of the element in which the mesh size is determined. In addition, in general lines, it should be mentioned that in places where the concentration of stress is greater, it is also convenient to increase the density of the mesh. Figure 14 shows the meshing result that is generated automatically by Autodesk Inventor Professional.

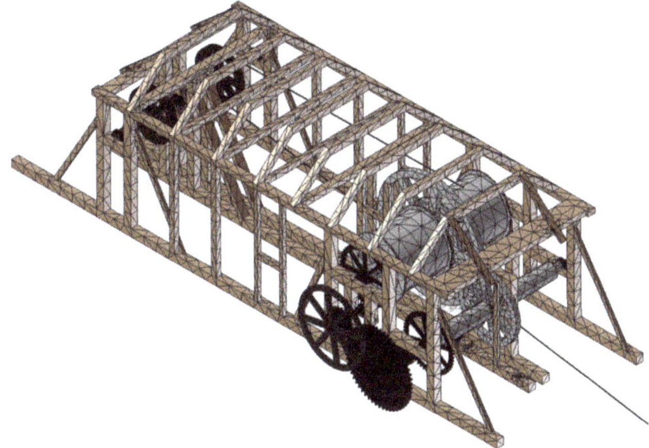

Figure 14. Automatic mesh obtained from the Hay inclined plane.

In the present research, it was also necessary to modify the discretization in some places, performing a refinement of the mesh. Specifically, this was done in the contacts between the teeth of adjacent gears (Figure 15) and on the surfaces of the ends of the axles in contact with their supports (Figure 16). In the first case, the reason for this intervention was the complicated geometry of the tooth of the gear, while in the second case, the reason lay in the concentration of the stresses.

Figure 15. Refinement of the mesh in the teeth of the gears.

Figure 16. Refinement of the mesh at the ends of the shafts in contact with the supports.

To conclude this section, it must be emphasized that for static analysis, it is necessary to establish a convergence criterion since iterative processes are used in this analysis. Thus, the result obtained is compared with the previous analysis and when it differs by less than 5% the process stops. In this research, and taking into account the computational resources available, the analysis used a mesh of 1,028,444 elements and 1,660,194 nodes in the blocking situation, and in the situation of emergency braking a mesh of 1,417,521 elements and 2,237,388 nodes.

3. Results and Discussion

On the interpretation of the results, it is convenient to point out that von Mises stresses were used instead of principal stresses. Although these show somewhat lower values than the principal stresses for the elements made of wood, it does not mean that these values are not valid, but they provide important indicative values, which, moreover, will not be too far from the value maximum stress in the most unfavorable direction.

Furthermore, the main direction of the wooden element (its greater length) is the direction parallel to the grain (the most resistant), and the engineer thought of wooden elements to work in compression. This indicative data, which shows the calculation of the von Mises stress on the wooden elements, is far from the elastic limit of the structure, whereas it shows large concentrations of stress on metal (isotropic) elements, and in this case, the von Mises stress is a precise datum of the load which the mechanism supports.

3.1. Modal Analysis

Before carrying out the static analysis, it is necessary to check previously, by means of a modal analysis, whether the engine behaves like a rigid solid. If it had such behavior, the software would provide erroneous results.

To do this, the model is subjected to a series of vibrations at different frequencies, which produces some deformations so that at high frequencies, elements appear that have a deformation greater than the reference deformation adopted by the software. Autodesk Inventor Professional provides the 8 lowest frequencies at which any part of the model is deformed, and if those frequencies are close to 0 Hz, it means that the element behaves like a rigid solid, and therefore, it would not make sense to perform a static analysis.

For the blockage situation, the following frequencies were obtained: F1: 2.00 Hz, F2: 2.81 Hz, F3: 2.83 Hz, F4: 5.78 Hz, F5: 6.25 Hz, F6: 8.11 Hz, F7: 9.00 Hz and F8: 9.04 Hz. This means that, in this situation and as expected, the model behaves statically and, therefore, the study can proceed.

For the emergency braking situation, the frequencies were different since some elements do not come into action. The results obtained were: F1: 3.39 Hz, F2: 4.25 Hz, F3: 6.37 Hz, F4: 6.42 Hz,

F5: 6.71 Hz, F6: 9.08 Hz, F7: 9.26 Hz, F8: 9.46 Hz. Thus, the modal analyses of both situations were very similar without loads.

On the other hand, as explained previously, the mesh size affects the exact places where stresses are concentrated, so the software allows the possibility of establishing a study on the convergence of the results by iterating the results in the regions that are considered more convenient. Therefore, for the static analysis a maximum of 10 cycles of iteration has been established, as long as the results vary more than 5% in each iteration. If the results in a region are less than that 5%, it stops iterating. In the same way, the greater the maximum number of iterations and the smaller the variation between results, evidently, the greater the computational requirements and the more prolonged the simulation time. In any case, this study is necessary in order to obtain a reliable result.

The results of the convergence curves for the two situations studied are shown in Figure 17. In the blockage situation the convergence rate (0.035%) is reached at the fifth iteration (Figure 17a), and in the emergency braking situation, the convergence rate is lower (0.335%), reached at the third iteration (Figure 17b). Therefore, the evolution of both graphs allows us to determine that the results obtained is very reliable.

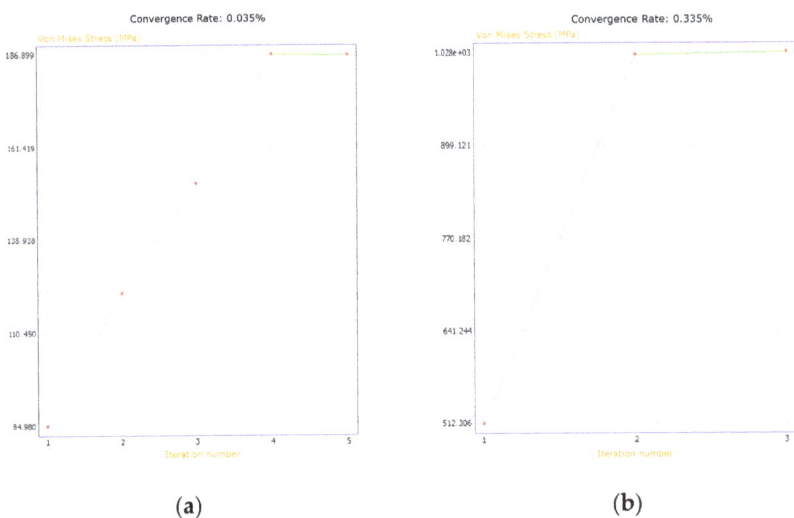

Figure 17. Convergence curve: (**a**) blockage situation and (**b**) braking situation.

3.2. Static Analysis

As can be supposed, the results of the static analysis also vary depending on the situation in which the mechanism is studied.

In the blockage situation, the maximum stresses are located on the supports of the horizontal axles. At these points, the von Mises stress reaches a maximum at the insertion of the DD axle with the inertia flywheel with a value of 186.9 MPa (Figure 18a). Similarly, for the rest of the contacts between the axles and the supports, there also appear high stress values: the support of the AA axle is subjected to a stress of 169.5 MPa (Figure 19a), and the support of the intermediate BB axle to a stress of 77.8 MPa (Figure 20a). Outside of these points, stresses are generally low.

In the braking situation, the maximum stress is located at the point of attachment of the support bar of the first brake, with a value of 1025 MPa (Figure 18b). This stress, as was observed in the previous sections, could not be achieved, since the operator would need to use a force equivalent to 2 tons to be able to stop the loaded tugboat. Therefore, the support bar of the first brake would break. However, it should be emphasized that the rest of the structural elements would be subject to reasonable stress. Thus, the second highest stress value would be reached in the support of the bar with a value of 390.7 MPa (Figure 19b), and the third highest value (287.1 MPa), would be reached at one part of the drive shaft, and would be located at the point of contact between the shaft of the largest gear of the DD shaft and that of the BB shaft gear (Figure 20b).

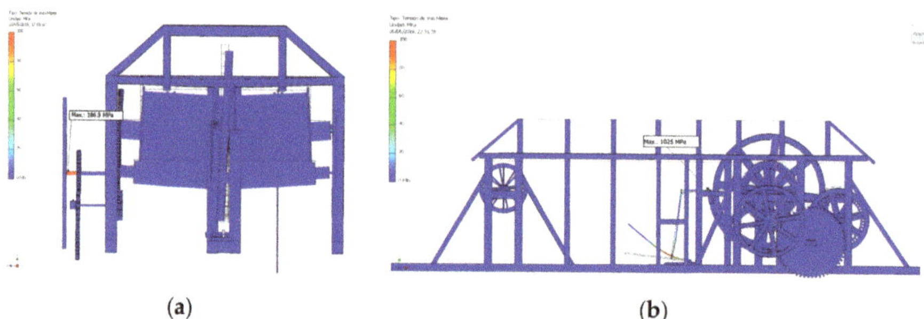

(a)   (b)

Figure 18. Location of the highest von Mises stress: (**a**) blockage situation and (**b**) braking situation.

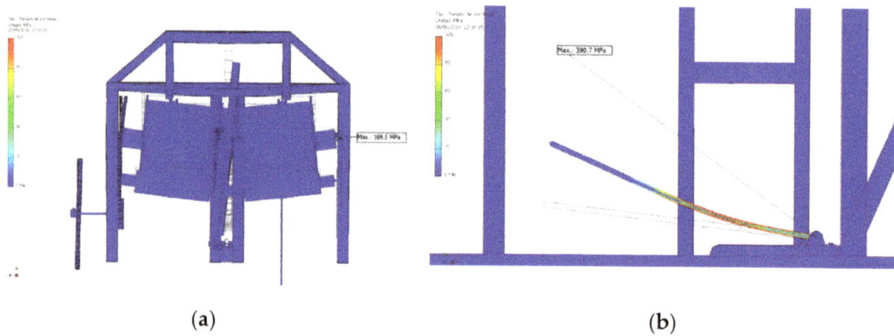

(a)   (b)

Figure 19. Location of the second highest von Mises stress: (**a**) blocking situation and (**b**) braking situation.

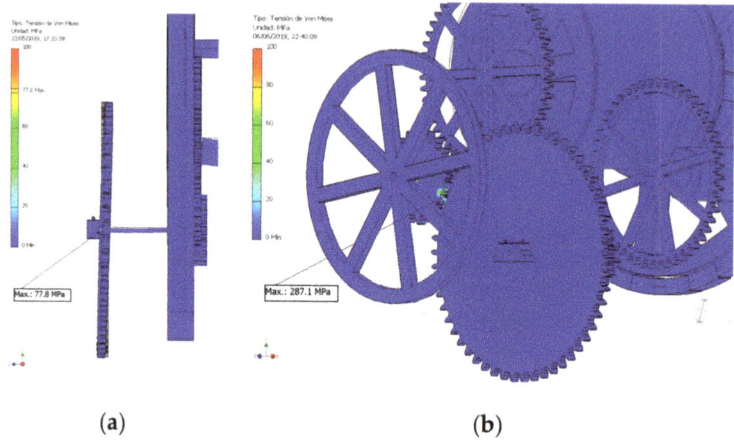

(a)   (b)

Figure 20. Location of the third highest von Mises stress: (**a**) blocking situation and (**b**) braking situation.

Another result of the static analysis is that relative to the safety coefficient. The safety coefficient at a given point in the model is calculated as the ratio between the von Mises stress and the elastic limit of a material for that point. In other words, the safety coefficient shows graphically whether the material from which a piece is made is capable of supporting a certain stress or not. In principle, a safety factor close to one unit would indicate that the material is working very close to its elastic limit; if it is less than one, it would indicate that the material exceeds this limit and would break, and if it is greater than one, it would indicate that it works without problems when subjected to said stress.

Nowadays, the materials and their elastic limits are much better known than at the end of the 18th century, and for this reason, engineers aim for the elements to work within a range of safety coefficient of 2 to 4 units. At the time of the *Hay inclined plane,* on the contrary, machines tended to be oversized, since those limits were unknown. For this reason, it is not surprising that, in general, almost the entire model works well above the elastic limit of the materials.

In the blockage situation, it can be observed, however, that the axle supports work with a value of 3.51 (Figure 21a). These elements have in common that they are in contact with the most stressed elements, and therefore, William Reynolds used wooden supports, because they are elements that can be easily replaced, since there was much wear due to the excessive friction of the axles (the bearing had not yet been invented).

However, in the braking situation, the point of maximum stress (point of union of the support bar of the first brake) coincides with the one with the lowest safety factor, with a value of 0.74 (Figure 21b). Outside this bar, the points that have the lowest safety coefficient are located at the joints between the links that form the brake, with a value of 1.7 in the tenth link (Figure 22b). The rest of the links have somewhat higher values, also around 2.

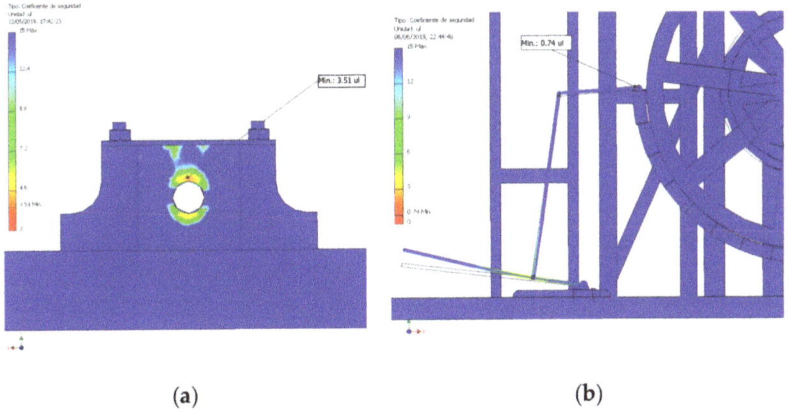

(a)   (b)

Figure 21. Location of the lowest safety coefficient: (**a**) blockage situation and (**b**) braking situation.

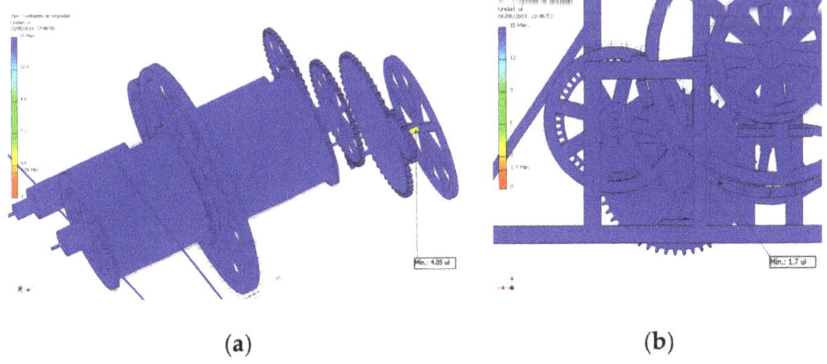

(a)   (b)

Figure 22. Location of the second lowest safety coefficient: (**a**) blockage situation and (**b**) braking situation.

Also, in the blockage situation, it must be mentioned that the union between the inertia flywheel and the axle is another delicate place. The flywheel, which is the element for which movement had been restricted, suffers an important torque in its axle. However, the safety factor of the highest stress point is 4.05, and therefore it is well dimensioned, but it is by far the point of the transmission shaft that works with the lowest safety coefficients (Figure 22a).

Finally, and also for the blockage situation, the distal support of the AA axle is shown. In this place, there is the element with the third lowest safety coefficient with a value of 4.13 (Figure 23), but already forming part of the set of oversized elements, as happens with the rest of the pieces of the mechanism.

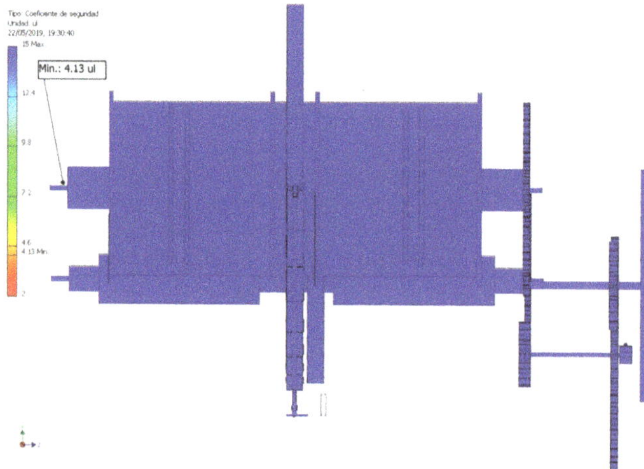

Figure 23. Third lowest safety coefficient in the blockage situation.

Another aspect that static analysis studies is that of the deformations suffered by the various elements, as these play an important role in the proper functioning of the mechanism. Thus, an excessively deformed element can generate problems for the correct performance of its function, affecting the proximate elements. For example, those elements inserted in guides, crossed by bolts or those that need to keep their contacts well defined in order to gear without gaps are very sensitive to deformations.

In the blockage situation, the maximum equivalent deformation, located at the junction of the inertia flywheel with the DD axle has a value of 0.13% with respect to the size of the element, and, therefore, is negligible (Figure 24a).

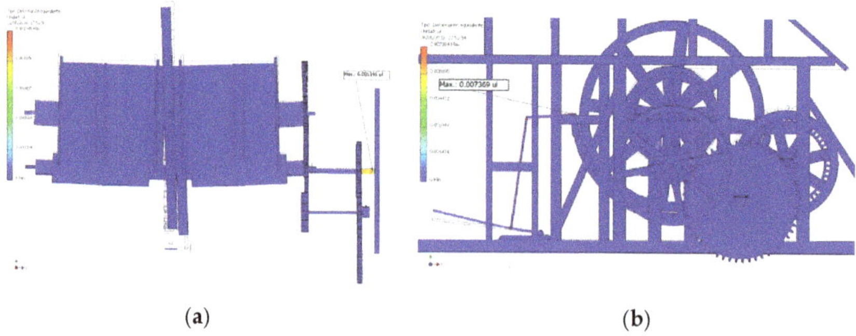

(a)   (b)

Figure 24. Location of the highest deformation: (**a**) blockage situation and (**b**) braking situation.

However, in the braking situation (Figure 24b), the maximum equivalent deformation, located at the insertion of the braking bar with the support of the first link of the brake and according to the point of maximum stress, is somewhat greater, with a value 0.74%, although very far from representing a danger for the operation of the mechanism.

Finally, the displacements are analyzed. For reasons similar to those mentioned in the previous section, the elements subjected to a stress point can present important displacements with respect to their usual working position. However, and in general, it can be seen that most of the elements hardly suffer any displacement when subjected to normal loads.

In the blockage situation, the piece subjected to the greatest displacement, with a value of 22.98 mm, is the inertia flywheel (Figure 25a). Although this is an acceptable displacement, this data indicates again that this piece is affected by the stopping of the transmission shaft and suffers like no other the consequences of this effort. The second point where the displacement is greatest is located at a point in the structure (Figure 26a) with a value of 6.27 mm, which indicates that the displacements in the rest of the elements of the transmission shaft are negligible, and that the structure works correctly.

In the braking situation, the end of the actuated lever is the point registering the greatest displacement with a value of 247.1 mm (Figure 25b). It is a solid bar of cast iron 2 m in length, so although such a displacement is not acceptable, it is understandable given the characteristics of the force that causes it (close to 22 kN). The second maximum displacement is located at the inertia flywheel, as also happens in the blockage situation, with a value of 10.28 mm (Figure 26b). The rest of the elements present negligible displacements.

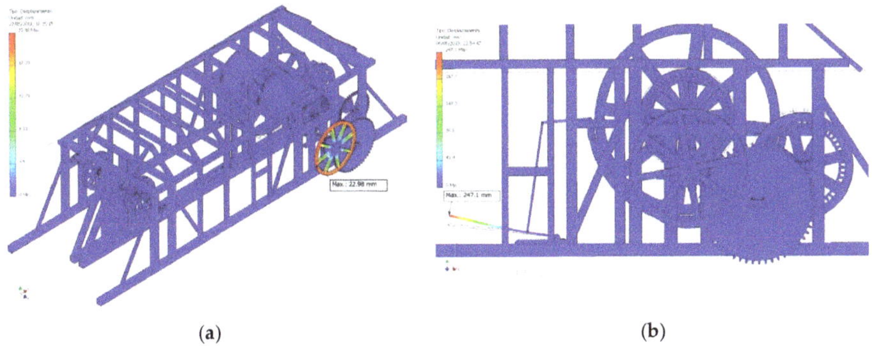

(**a**) (**b**)

Figure 25. Location of the highest displacement: (**a**) blockage situation and (**b**) braking situation.

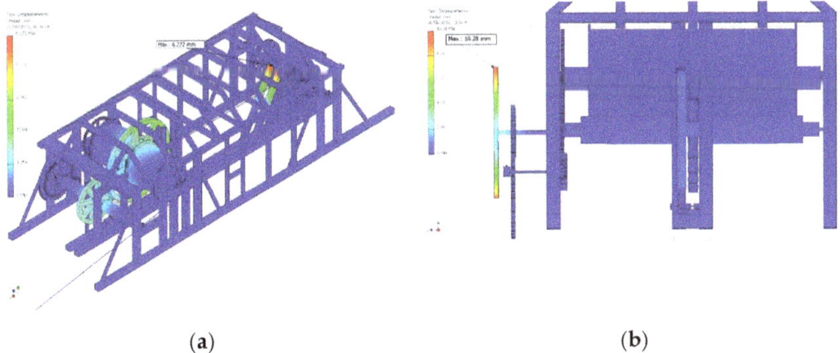

(**a**) (**b**)

Figure 26. Location of the second highest displacement: (**a**) blockage situation and (**b**) braking situation.

4. Conclusions

- In the present research, static analysis by FEM is shown, applied to the 3D model of the *Hay inclined plane* located in Coalbrookdale (Shropshire, England), the work of the English engineer William Reynolds, just as it worked a few years after its inauguration in 1792. This 3D model was developed based on the work of the Spanish engineer Agustín de Betancourt, who published a report about the method used by Reynolds to negotiate uneven channels using inclined planes. For these tasks, Autodesk Inventor Professional software was used.

- Static analysis was used to study the behavior of the engine in the two most unfavorable situations: the first one, in which a inertia flywheel blockage due to a problem in the steam engine is simulated, and a second one in which emergency braking of a loaded tugboat is required by the disconnection between the inertia flywheel and the steam engine.

- The results of the static analysis were very different, depending on the situation analyzed. For the blockage situation, a maximum stress of 186.9 MPa was obtained at the insertion point of the inertia flywheel with the DD axle. In addition, the wooden support of said DD axle had the lowest safety coefficient with a value of 3.51, which indicates that it is well dimensioned. The same DD axle would have a negligible maximum deformation of 0.13%, and the maximum displacement is located at the inertia flywheel, with a value of 22.98 mm. Similarly, it was observed that the rest of the elements always had values on the side of safety, so for this situation, it can be said that the mechanism is oversized; that is, the materials work far below their elastic limit, just as happened in the designs of the time.

- However, the results were different in the emergency braking situation. On the one hand, as stated above, this second situation, although it would be highly advisable in the face of an emergency situation, was impossible to resolve in practice. That is, unless there was a mechanical system for actuating the lever (which is not described in the written account, but would certainly have existed), it would be impossible to exert sufficient braking pressure using human means. As was seen, the force that it would be necessary to exert on the brake bar to stop the mechanism was almost 22 kN, very far beyond the capacity of a human operator. In any case, it was considered useful to approach the study of this hypothetical situation as a method for evaluating the mechanism structurally.

- Also, the static analysis of this emergency braking situation was very revealing. For this situation, a maximum stress of 1025 MPa is reached at the point of insertion of the brake bar and the first link of the brake, and since the elastic limit of cast iron is 758 MPa, at that point the bar would break. In addition, the security coefficient at this point is 0.74, confirming, therefore, the invalidity of its dimensions. However, the analysis is surprising for two reasons: on the one hand, a low security coefficient is not necessarily excessive, so a simple change in the dimensions of the bar would be enough for this element to fall within the safety range; and on the other hand, it is striking that it is the only point where these values are exceeded. The second maximum stress is located at the insertion point of the braking bar with its support presenting an acceptable value of 390.7 MPa, and the second point with the lowest safety coefficient corresponds to the links that join the brake parts with a value of 1.7, which falls within the safety range. As indicated above, although this emergency braking situation could not occur manually, it is verified that for such a large braking stress the structure would function correctly except in the dimensioning of the braking bar. In fact, this bar is the element where the greatest equivalent deformation and the largest displacement within the mechanism is located, with values of 0.73% and 247.1 mm, respectively. The latter value would suggest the need for a resizing of the bar.

- All these data show that by the time it was built, the invention was correctly sized. The hypothesis of work in the situation of emergency braking indicates, on the other hand, that it would be difficult to stop a loaded tugboat using the axle brake (which would entail a certain danger for loads and people), but at the same time, it shows that, despite this, the general structure (except for the braking bar) was well dimensioned. The *Hay inclined plane* was in operation from 1792 to 1944, and this fact corroborates all the results shown here. At the same time, the present research emphasizes that, from the point of view of mechanical design, the invention of William Reynolds has all the qualities to qualify it as a 'masterpiece'.

Author Contributions: Formal analysis, J.I.R.-S. and E.D.l.M.-D.l.F.; Funding acquisition, J.I.R.-S.; Investigation, J.I.R.-S. and E.D.l.M.-D.l.F.; Methodology, J.I.R.-S. and E.D.l.M.-D.l.F.; Project administration, J.I.R.-S.; Supervision, J.I.R.-S.; Validation, E.D.l.M.-D.l.F.; Visualization, J.I.R.-S.; Writing—original draft, J.I.R.-S. and E.D.l.M.-D.l.F.; Writing—review & editing, J.I.R.-S. and E.D.l.M.-D.l.F.

Funding: This research was funded by the Spanish Ministry of Economic Affairs and Competitiveness (MINECO), under the Spanish Plan of Scientific and Technical Research and Innovation (2013–2016) and European Fund Regional Development (EFRD) under grant number [HAR2015-63503-P].

Acknowledgments: We are very grateful to the Fundación Canaria Orotava de Historia de la Ciencia for permission to use the material of Project Betancourt available at their website. Also, we sincerely appreciate the work of the reviewers of this manuscript.

Conflicts of Interest: The authors declare no conflict of interest.

References

1. Muñoz Bravo, J. *Biografía Cronológica de Don Agustín de Betancourt y Molina en el 250 Aniversario de su Nacimiento*; Acciona Infraestructuras: Murcia, Spain, 2008. (In Spanish)
2. Martín Medina, A. *Agustín de Betancourt y Molina*; Dykinson: Madrid, Spain, 2006. (In Spanish)
3. Cioranescu, A. *Agustín de Betancourt: Su Obra Técnica y Científica*; Instituto de Estudios Canarios: La Laguna de Tenerife, Spain, 1965. (In Spanish)
4. Rojas-Sola, J.I.; Galán-Moral, B.; De la Morena-De la Fuente, E. Agustín de Betancourt's double-acting steam engine: Geometric modeling and virtual reconstruction. *Symmetry* **2018**, *10*, 351. [CrossRef]
5. Rojas-Sola, J.I.; De la Morena-de la Fuente, E. Digital 3D reconstruction of Agustín de Betancourt's historical heritage: The dredging machine of the port of Krondstadt. *Virtual Archaeol. Rev.* **2018**, *9*, 44–56. [CrossRef]
6. Rojas-Sola, J.I.; De la Morena-de la Fuente, E. Geometric modeling of the machine for cutting cane and other aquatic plants in navigable waterways by Agustín de Betancourt y Molina. *Technologies* **2018**, *6*, 1. [CrossRef]
7. Rojas-Sola, J.I.; De la Morena-de la Fuente, E. Agustín de Betancourt's wind machine for draining marshy ground: Approach to its geometric modeling with Autodesk Inventor Professional. *Technologies* **2017**, *5*, 2. [CrossRef]
8. Rojas-Sola, J.I.; De la Morena-De la Fuente, E. Agustín de Betancourt's mechanical dredger in the port of Kronstadt: Analysis through Computer-Aided Engineering. *Appl. Sci.* **2018**, *8*, 1338. [CrossRef]
9. Rojas-Sola, J.I.; De la Morena-De la Fuente, E. Agustín de Betancourt's double-acting steam engine: Analysis through Computer-Aided Engineering. *Appl. Sci.* **2018**, *8*, 2309. [CrossRef]
10. Rojas-Sola, J.I.; De la Morena-De la Fuente, E. Agustín de Betancourt's wind machine for draining marshy ground: Analysis of its Construction through Computer-Aided Engineering. *Inf. Constr.* **2018**, *70*, e236. [CrossRef]
11. Rojas-Sola, J.I.; De la Morena-De la Fuente, E. Agustín de Betancourt's piston lock: Analysis of its Construction through Computer-Aided Engineering. *Inf. Constr.* **2019**, *71*, e286. [CrossRef]
12. Rojas-Sola, J.I.; De la Morena-De la Fuente, E. Agustín de Betancourt mil for grinding flint: Analysis by Computer-Aided Engineering. *Dyna* **2018**, *93*, 165–169. [CrossRef]
13. Rojas-Sola, J.I.; De la Morena-De la Fuente, E. The Hay inclined plane in Cooalbrookdale (Shropshire, England): Geometric modeling and virtual reconstruction. *Symmetry* **2019**, *11*, 589. [CrossRef]
14. Hadfield, C. *Thomas Telford's Temptation*; M. & M. Baldwin: Cleobury Mortimer, UK, 1993.
15. Betancourt, A. Dessin de la Machine Pour Faire Monter et Descendre les Bateaux d'un Canal Inferieur a un Superieur et Reciproquement Sur Deux Plans Inclines. Available online: http://fundacionorotava.org/pynakes/lise/betan_dessi_fr_01_179X/0/?zoom=large (accessed on 15 August 2019).
16. Betancourt, A. Mémoire Sur un Nouveau Système de Navigation Intérieure. Available online: http://fundacionorotava.es/pynakes/lise/betan_memoi_fr_01_1807 (accessed on 15 August 2019).
17. Fulton, R. *A Treatise on the Improvement of Canal Navigation*; I. and J. Taylor: London, UK, 1796.

18. Hay Inclined Plane (from the Bottom Canal) in 1879. Available online: https://captainahabswaterytales.blogspot.com/search?q=hay+plane (accessed on 15 August 2019).

19. Hur, D.J.; Know, S. Fatigue analysis of greenhouse structure under wind load and self-weight. *Appl. Sci.* **2017**, *7*, 1274. [CrossRef]

20. Li, J.B.; Gao, X.; Fu, X.A.; Wu, C.L.; Lin, G. A nonlinear crack model for concrete structure based on an extended scaled boundary finite element method. *Appl. Sci.* **2018**, *8*, 1067. [CrossRef]

21. Fundación Canaria Orotava de Historia de la Ciencia. Available online: http://fundacionorotava.org (accessed on 15 August 2019).

22. Proyecto Betancourt. Available online: http://fundacionorotava.es/betancourt (accessed on 15 August 2019).

23. Shih, R.H. *Parametric Modeling with Autodesk Inventor 2016*; SDC Publications: Mission, KS, USA, 2015.

 *applied*
sciences

Article

Computer Aided Design to Produce High-Detail Models through Low Cost Digital Fabrication for the Conservation of Aerospace Heritage

Jose Luis Saorín [1], Vicente Lopez-Chao [1,*], Jorge de la Torre-Cantero [1] and Manuel Drago Díaz-Alemán [2]

[1] Engineering Graphics Area, University of La Laguna, 38204 Tenerife, Spain; jlsaorin@ull.edu.es (J.L.S.); jcantero@ull.edu.es (J.d.l.T.-C.)
[2] Department of Fine Arts, University of La Laguna, 38204 Tenerife, Spain; madradi@ull.edu.es
* Correspondence: vlopezch@ull.edu.es

Received: 7 May 2019; Accepted: 3 June 2019; Published: 6 June 2019

Abstract: Aerospace heritage requires tools that allow its transfer and conservation beyond photographs and texts. The complexity of these engineering projects can be collected through digital graphic representation. Nevertheless, physical scale models provide additional information of high value when they involve full detailed information, for which the model in engineering was normally one more product of the manufacturing process, which entails a high cost. However, the standardization of digital fabrication allows the manufacture of high-detail models at low cost. For this reason, in this paper a case study of the graphic reengineering and planning stages for digital fabrication of a full-scale high-detail model (HDM) of the spatial instrument of the European Space Agency, named the Solar Orbiter mission Polarimetric and Helioseismic Imager (SO/PHI), is presented. After the analysis of this experience, seven stages of planning and graphic reengineering are proposed through collaborative work for the low cost digital manufacture of HDMs.

Keywords: engineering graphics; digital manufacture; 3D printing; computer-aided design; Autodesk Fusion 360; aerospace heritage

---

1. Introduction

The Design and Digital Manufacturing Laboratory of the University of La Laguna (Fab Lab ULL) has worked in the digital fabrication of a full-scale high-detail model (HDM) of the spatial instrument of the European Space Agency, named the Solar Orbiter mission Polarimetric and Helioseismic Imager (SO/PHI) [1]. This sort of physical replica entails deep information, which is essential for the preservation and transfer of aerospace heritage since projects launched into space are normally designed in order to not return to Earth.

The manufacturing of the HDM of the SO/PHI instrument was carried out under an approach that combined digital manufacturing techniques (low cost 3D printers, a CNC milling machine, and a laser cutting machine) and the use of a 3D computer-aided collaborative modeling software. The aim of this paper is to provide stages in the graphic engineering procedure for the digital fabrication of challenging projects, which require high detail through the resolution of a case study: the manufacture of a full-scale HDM of the spatial instrument PHI for the Solar Orbiter mission of the European Space Agency.

Rapid prototyping has made it possible to print 3D physical prototypes from CAD files [2]. However, this process is not always fast (since the model is not a single piece, and the printing process is not instantaneous). This lack of immediacy is plausible in architecture environments, where reduced scale models are indispensable and geometric simplification processes are required since buildings cannot be uniformly scaled [3].

3D digital printing precision and other fabrication variabilities have been a focus of interest in the scientific literature, including studies that compare the difference between the CAD model and the physical model [4]. In addition, other issues have dealt with the use of different digital manufacturing techniques with materials such as concrete [5] or the study of numerical models to analyze the behavior of timber folded surface structures using semi-rigid multiple tab and slot joints [6].

Nevertheless, there are many other challenges [7] when we face manufacturing a full-scale HDM of an engineering project. Although details are easily achieved in digital modeling, physical models provide important additional information for the conservation of engineering heritage, such as SO/PHI. Therefore, this article proposes a procedure to recreate engineering projects without the need for geometric simplifications through the use of low-cost digital fabrication methods.

1.1. Replicas for Aerospace Industrial Heritage

Heritage conservation awareness has increased and therefore, engineering graphics literature has recognized more and more research in different areas of heritage, whether industrial [8], sculptural [9], fossil [10], or architectural [11]. The term industrial heritage has been related to remains of constructions or inventions. However, in certain areas of engineering that possibility does not come to be, as it is in the aerospace sector. Therefore, space engineering projects has become critical evidence of the development of humanity, and consequently they remain part of our heritage [12]. For this reason, some researchers have demanded the need to act in the cataloguing and storage of aerospace heritage, indicating that published data are insufficient since they simply state the launch date, the name of the satellite, and the orbit [13].

Graphic representation is one of the tools that can contain and relate more information. In this sense, there are investigations focused on the use of digital representation such as the use of photogrammetry for the generation of three-dimensional modeling [14], augmented reality to visualize the industrial heritage [15], the application of BIM techniques for the representation of architectural heritage with a high level of detail [16] and the use of simplified models in educational environments to promote cultural heritage [17], including the supply of NASA stl files [18].

Photogrammetry and BIM can make a digital 3D model, the first one is faster has lower precision than the second one. Once the 3D model is created, it can be viewed digitally (augmented reality) or physically (scale model). The first has the level of detail that has been granted in digital modeling, while physical models usually entails geometric simplifications of the engineering project due to manufacturing or financial limitations. However, physical models allow a person to approach them in their real environment, so they can manipulate and understand them through the sense of touch. Meanwhile, the digital environment uses zoom or rotation tools to approach the pieces or rotate them, which can generate a distorted view of reality.

Although the 3D modeling programs have increased their use in engineering environments for the visualization and evaluation of the designed elements [19], tangible models are still indispensable because the physical interaction with the model helps to understand it, evaluate it and detect design problems [20]. The additional information incorporated in the models has been a focus of analysis in historical research [21]. In addition, it allows a more direct and personal interaction and visualization by investors, clients, designers, and other professionals who may be involved in the project.

1.2. Scale Models: Architecture vs. Engineering

The model has been a vital tool for the design of architectural projects, since it was necessary to understand and visualize large-scale buildings in different stages in order to minimize and prevent problems that might arise. The feasibility of its elaboration was based on the use of straight geometries and orthogonal angles, which can reduce three dimensions to two dimensions, and the need to simplify details in order to address reducing scale models.

However, engineering projects need to reproduce complex geometries with high detail, such as the engine of a car. For this reason, scale models in this area have normally corresponded to industrial objects, made with the same process, cost, and materials as the original one.

Likewise, scale models of small elements have traditionally been made enlarging scale (instead of full scale), which has not allowed to visualize them in the context of the project. Currently, this precision can be achieved with low cost 3D imaging, by reaching resolutions up to 20 μm in PLA (polylactic acid by fused deposition modeling) and 0.01 μm in photosensitive resin (by digital light processing—DLP). Therefore, they become a fundamental tool for the development of full scale HDMs.

In architecture and design, prototyping has been conceived as a design method to study and test how a new product is going to be used, and how it will look [19]. This concept has conceived utility in models beyond the final result, so that there are different categories for the objectives of each phase of the project design. These categories can be classified as follows [22]:

- Soft model: Modeled by hand and allow exploration, and evaluate the size, proportion, and shape of concepts and ideas.
- Hard model: Usually made of wood, plastic, metal, or dense foam. It is not technically functional, but it is presented as a replica very close to the final design.
- Presentation model: With every detail of composition, in which the components have been simplified to save time.
- Prototype: High quality and functionality produced to exhibit a design solution.

Nonetheless, technological progress has generated changes in the manufacturing process of scale models, from conventional model making to methods such as rapid prototyping. The first method involves creative approaches and is more efficient in economic terms, while the second is an automated process through a digital model of the project, and higher cost. Thus, standardization of the use of low-cost 3D printers allows complex geometries and a high level of detail to be created whether both model manufacturing methods are combined.

1.3. Collaborative Working Environment in Engineering

Collaborative work has been fundamental to solve engineering challenges. In this sense, cloud storage services, collaborative environments, and social networks offer advantages and limitations [23] that despite its usefulness to communicate, write, and present results, do not allow 3D modeling.

In recent decades, engineering software companies have focused on solving this difficulty by offering high-cost collaborative work platforms (Autodesk Vault, PTC PLM Cloud, or 3D Experience). In this context, Fusion 360 from Autodesk emerges, a collaborative 3D modeling environment that is free for educational and for start-ups generating less than $100,000/year in total revenue or wholly non-commercial hobbyist users [24]. This software is easy to use, integrates several assisted design environments, and also has a smartphone application for viewing modeling, as well as offering a public web link to visualize the model in the browser [25].

In this article, the case study of the manufacture of a replica of the spatial instrument SO/PHI is provided. The incorporation of low-cost digital fabrication machines (below $2500) is proposed to solve the traditional characteristics in the manufacture of scale models such as geometric simplification and the high economic cost of detailed prototypes [20,22]. In addition, this approach aims to generate a framework of procedures and techniques of engineering graphics for the conservation and transfer of industrial heritage.

2. Materials and Methods

2.1. PHI Instrument and Solar Orbiter Mission

The Solar Orbiter mission, developed by the European Space Agency (ESA) in collaboration with NASA, focuses on the study of the Sun and magnetic activity in the heliosphere and will be able to obtain unique information to help understand the operation of this star and even predict its behavior. One of its devices, the PHI instrument (composed of about 1500 pieces) is the largest and perhaps the most complex. It will provide high-resolution and full-disk measurements of the photospheric vector magnetic field and line-of-sight (LOS) velocity as well as the continuum intensity in the visible wavelength range.

Due to its complexity, this instrument has been developed by an international consortium involving Germany, Spain, and France among other countries. The coordination of the Spanish part is carried out by the Institute of Astrophysics of Andalusia (IAA-CSIC), the Institute of Astrophysics of the Canary Islands (IAC), the National Institute of Aerospace Technology (INTA), the University of Valencia, the University of Barcelona, and the Technical University of Madrid.

Fab Lab ULL has produced the two-dimensional graphic documentation and the full-scale high-detail model of the SO/PHI, in order to explain its morphology and disposition of subassemblies. This physical model allows the dissemination of the research carried out by the Spanish part and is intended to have the same level of detail as the original project.

2.2. Case Study Procedure

The project begins with the reception of the original digital model that contains all the information of almost 1500 pieces modeled in 3D in step format. From them, a process of graphic re-engineering is carried out to prepare the pieces for their manufacture. First, six teams prepare the plans for the manufacture of the model. Subsequently, two teams and two supervisors create the replica through additive techniques (3D printing in polylactic acid, and photosensitive resin) and subtractive techniques (CNC laser cutting and cutting in PVC and methacrylate).

The resolution of this problem, which enables the conservation and transmission of aerospace heritage, is presented as a case study to establish graphic engineering procedures that allow the creation of HDMs based on low-cost digital manufacturing.

2.3. Materials

2.3.1. Collaborative CAD Software

The process is conducted in Fab Lab ULL and Autodesk Fusion 360 is used. The choice of this parametric design tool lies in its viability to work collaboratively as it is developed in an online environment. It also offers a platform in the cloud of Autodesk for the management and visualization of 3D models called A360. This software is conceived for the manufacture of digital products, through 3D printing, CNC milling, or laser cutting.

This research relates the resolution of a HDM through collaborative work and with low cost digital fabrication machines. In this way, the process can be replicated by small companies that are dedicated to the field of modeling.

2.3.2. Manufacture Means

Three low-cost 3D printers are used in the project: Two for printing by fused deposition (PLA) and one for DLP digital light processing (photosensitive resin), a laser cutting machine, and a CNC milling machine. Table 1 shows the technical specifications of the 3D printers.

Table 1. Technical specifications of the 3D printers.

3D Printer	Prize (€)	Printing Material	Resolution (μm)	Print Volume (mm)
BQ witbox 2	<1400	PLA	up to 20	297 × 210 × 200
Wanhao duplicator 7	<750	ABS	up to 0.01	120.96 × 68.5 × 180

2.3.3. Materials for the Manufacture of the SO/PHI HDM

For the manufacture of the model, the following materials are used: PVC and methacrylate (for the largest bi-dimensional parts), PLA (polylactic acid—for the model elements that require greater complexity and detail), photosensitive resin (for those elements that require higher levels of precision), acrylic paint (applied with a spray gun to achieve a uniform and homogeneous result), aluminium rods, and steel sheets.

3. Results

3.1. Re-Engineering of the Original 3D Model

3.1.1. Identification and Organization of Sub-Assemblies

In a first phase, it is necessary to organize the project into smaller and more manageable parts. For this, the different subassemblies that make up the PHI instrument assembly are grouped and a workspace is created in A360 for the management of the files. This consists in the creation of a Fusion 360 project that is shared with all the Fab Lab ULL team members who worked on this digital model. Thus, each member has access to all the information from the first moment. Furthermore, each person has one or more sub-assemblies assigned.

The classification and labelling system of the 17 sub-assemblies that structure the SO/PHI is essential. For this, transparent plastic containers with labels and with the plans for their manufacture are prepared to store, in an orderly manner, all the pieces that were going to be printed of each subassembly (see Figure 1).

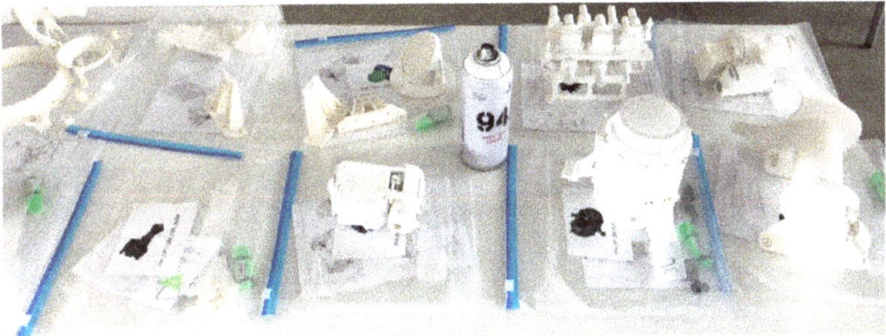

Figure 1. Classification and labelling of SO/PHI sub-assemblies.

3.1.2. Analysis of the Digital Files

The first problem consists in the management of the files and the organization of the subassemblies. Although the original model could have a certain organization by the different subassemblies, files in step format can lose part of the information when they are imported, so it is necessary to review and order their content.

File management begins with the detection of geometry problems. The fact that much of the digital model will be printed in 3D requires working with solid geometries, rather than surfaces. This factor is also important for the creation of plans since in Fusion 360 the elements modeled with surface geometries cannot be visualized in the 2D plans.

In the following pair of figures (Figure 2a,b) the received model (located on the left) and the model converted to solid are displayed. In the original file, solid geometries are perceived in the lower part of the subassembly, while in the rest, no cuts are generated because they are surface geometries. However, once the primitive geometry has been modified to solid geometry, the cut is visible throughout the sectioned part.

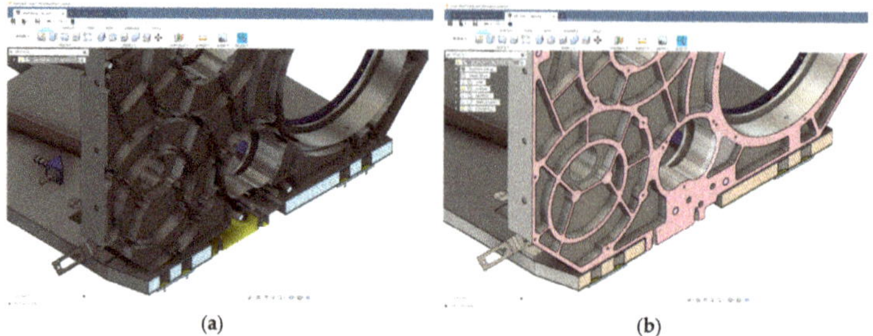

(a)   (b)

Figure 2. (**a**) Digital model with surface geometries. (**b**) Digital model with solid geometries.

3.1.3. Digital Model Repair

Once the detection and categorization of problems derived from the geometry is generated, it is required to correct the errors (overlaps, intersections...) and their subsequent transformation into solid geometry.

In the surface environment of Fusion 360, the stitch tool allows the surface to be sewn, and if it detects that the selection is a closed surface, it will convert it directly into a solid. If it were not a closed element, the interface highlights the perimeter lines of open surfaces in red (see Figure 3). In this case, the thicken tool can be applied to give thickness to the surface and convert it into a solid with the dimension that we have requested.

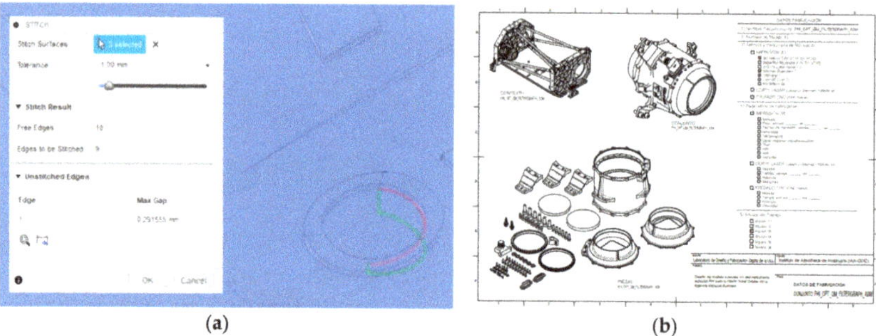

(a)   (b)

Figure 3. (**a**) Open surface detected through Stitch tool. (**b**) Subassembly 8 plan.

Once this process is completed for each subassembly, it is possible to start the creation of plans for manufacturing. In Figure 4b the planes of the Subassembly 8 are shown. Once all the surfaces are transformed into solids, all the parts (pieces, screws, and nuts) that make up each subassembly of the complete model can be quantified (Table 2).

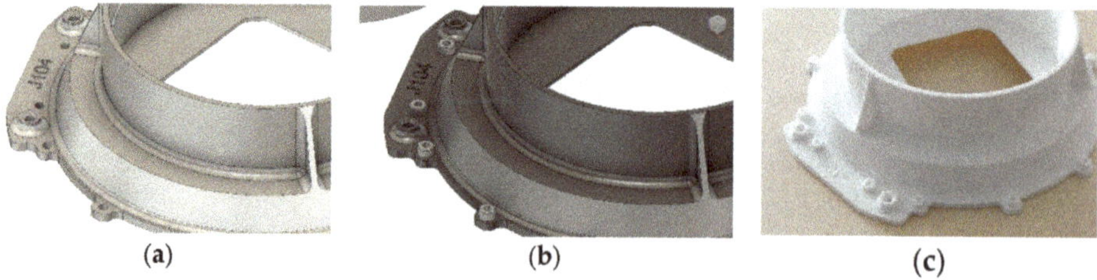

(a)   (b)   (c)

Figure 4. (**a**) Piece without screws in Fusion 360. (**b**) Piece with screws in Fusion 360. (**c**) Piece printed with screws.

Table 2. Frequency analysis of pieces that integrate the SO/PHI subassemblies.

	Pieces	Screws and Nuts	Total
Subassembly 1	25	58	
Subassembly 2	8	34	
Subassembly 3	10	37	
Subassembly 4	3	0	
Subassembly 5	3	36	
Subassembly 6	26	110	
Subassembly 7	3	22	
Subassembly 8	14	68	
Subassembly 9	11	52	
Subassembly 10	28	42	
Subassembly 11	2	7	
Subassembly 12	41	190	
Subassembly 13	8	0	
Subassembly 14	12	50	
Subassembly 15	49	107	
Subassembly 16	51	197	
Subassembly 17	19	146	
	313	1156	1469

3.1.4. Digital Model Restructuration to Optimize Their Manufacturing

It is essential to group geometries to optimize printing times. In particular, the screws and nuts (and other small elements) associated with each subassembly can be grouped in such a way that they can be printed in a single process. Screws and nuts can be embedded in subassembly parts since their mechanical operation is not necessary (i.e., Figure 4).

To shed light on the decrease in printing parts that this process entails, Table 3 presents data on the reduction of parts for Subassembly 8. In this case, the percentage of parts removal is 83. This situation is analyzed for each of the assembly subassemblies.

Table 3. Simplification of pieces of Subassembly 8.

Subassembly 8	Pieces
Original	82
Prepared for printing	14

In Figure 5, the geometric modeling of Subassembly 8 is exposed graphically, before and after restructuring the model for manufacturing.

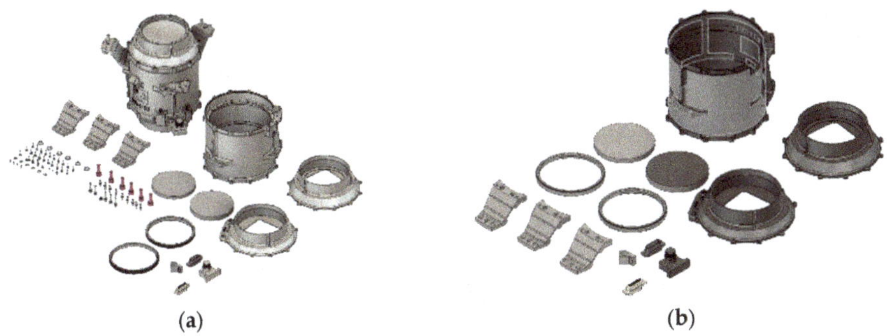

(a) (b)

Figure 5. (a) Original subassembly 8 (82 pieces). (b) Subassembly 8 prepared for manufacturing (14 pieces).

3.1.5. Adaptation of the Models to the Means of Manufacturing

The dimensions of the subassemblies are sometimes larger than the volume of the 3D printers. Therefore, it is necessary to divide the elements so that they adapt to our possibilities. In the upper left part of Figure 6, one of the parts into which the piece has been divided is located above the 3D printer work area.

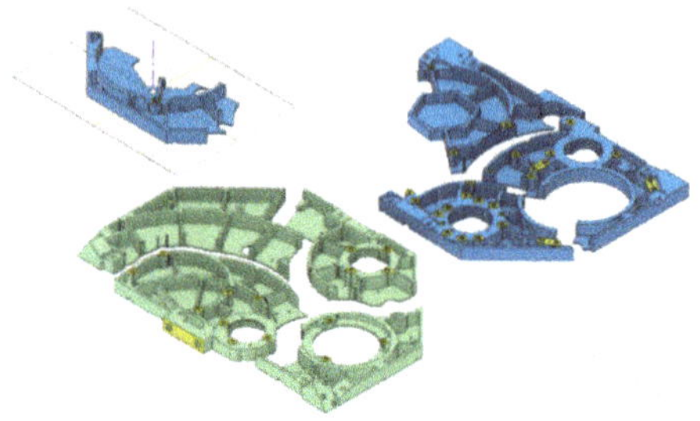

Figure 6. Piece divisions to fit in the 3D printer work area.

In addition, the material used in the engineering project is not the same used in the manufacture of the scale model. The geometry of the project is conditioned to its resistance in a specific material, so the change of material to print the pieces may require that geometric adaptations are generated. Some examples that need adaptations can be very thin fins or very long pieces. In both cases, it is necessary to add nerves or small tabs so that the pieces do not break when manufactured. Figure 7 shows an element that required the generation of internal reinforcements to increase its rigidity.

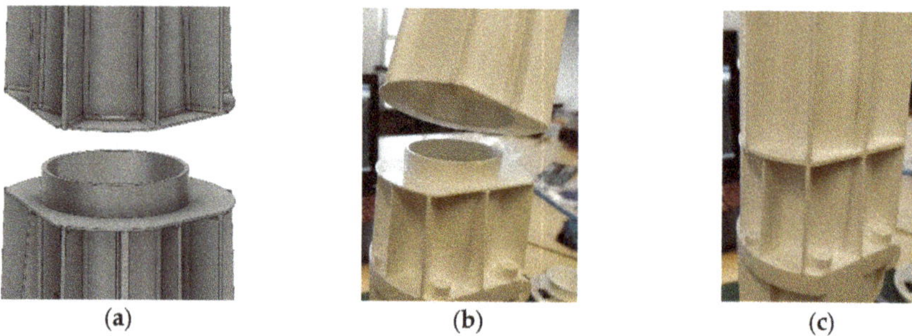

Figure 7. (a) View of the reinforcement connection between the two parts in which the piece has been divided. (b) Piece printed in two parts. (c) View of assembled printed pieces.

These geometry addition techniques may also be necessary with a shuttering function during 3D printing of the element to avoid buckling problems during printing, and which must then be removed from the final element.

3.1.6. Pre-Allocation of the Digital Models to the Manufacturing Means

Once the number of pieces to be printed has been reduced, it is necessary to assign which parts are printed with each 3D printer, depending on the need for resolution, or with the cutting machines, if they have laminar geometry. In order to reproduce very small pieces on a natural scale, photosensitive resin printing was used. This system has manufactured, among others, the connector of Subassembly 8 (one of the 14 pieces) due to the diameters of the pins. This analysis is performed for every part of each assembly. In Figure 8a the connector of the Subassembly 8 is exposed in context to the rest of the elements that form it and in Figure 8b a detail view of the connector.

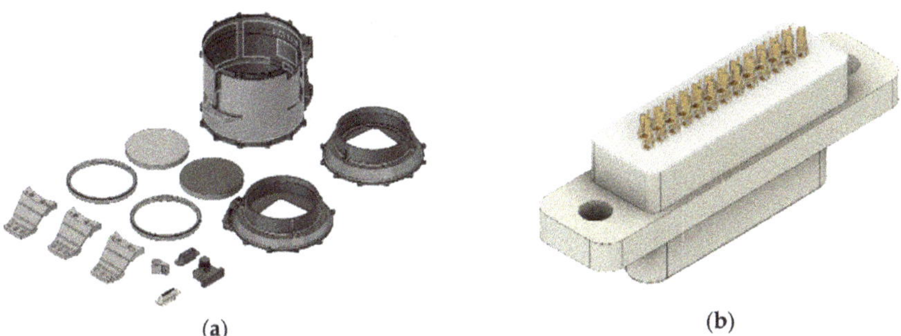

(a)   (b)

Figure 8. (**a**) Connector of Subassembly 8 in relation to the rest of its pieces. (**b**) Detail view of the connector.

In Figure 9a the printed connectors of the Subassembly 8 are exposed and in Figure 9b they are presented in relation to another piece of the same subassembly, where their dimensions can be compared with those of a hand. The pins of these connectors have diameters of 0.8 mm and thickness of 0.1 mm.

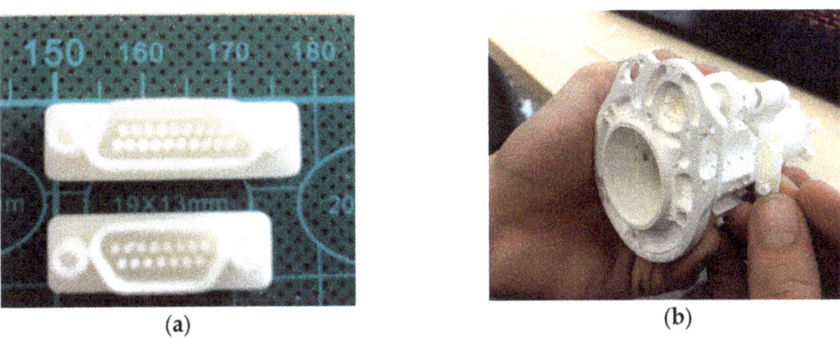

(a)   (b)

Figure 9. (**a**) Subassembly 8 connectors. (**b**) Connector and printed piece of Subassembly 8.

In SO/PHI, laminar pieces have been made by CNC milling and laser cutting. The support base of all the instrumentation is made in CNC milling, while the steel mirrors of the instrument and the methacrylate showcase are manufactured with a laser cutting machine. Figure 10 shows its large dimensions and its flat morphology, characteristics that respond to the methods of traditional model making in architecture. The reinforcing bars of the structure are manufactured by means of aluminum rods to which the printed ends in 3D have been added.

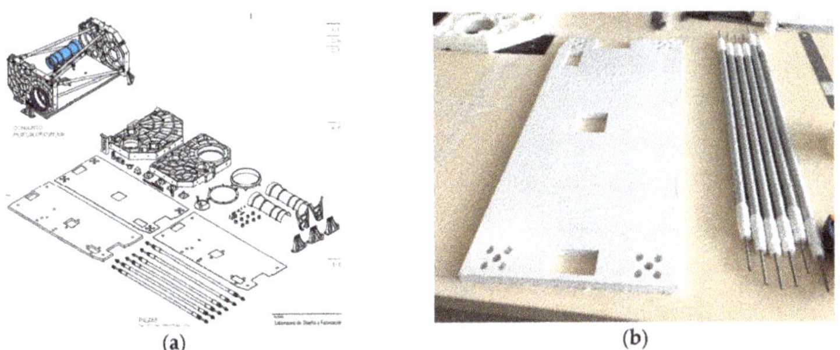

(a)   (b)

Figure 10. (**a**) Plane of the instrument base. (**b**) Base of the instrument and reinforcement rods.

3.1.7. Manufacturing and Finishing

Once the geometric modifications and the adjudication of the digital manufacturing machines have been made, it is necessary to make modifications in the printed elements in order to generate a uniform HDM in terms of texture and color. Figure 11a shows an example of joining pieces that had been printed in several parts, and in Figure 11b the process of painting to match the color.

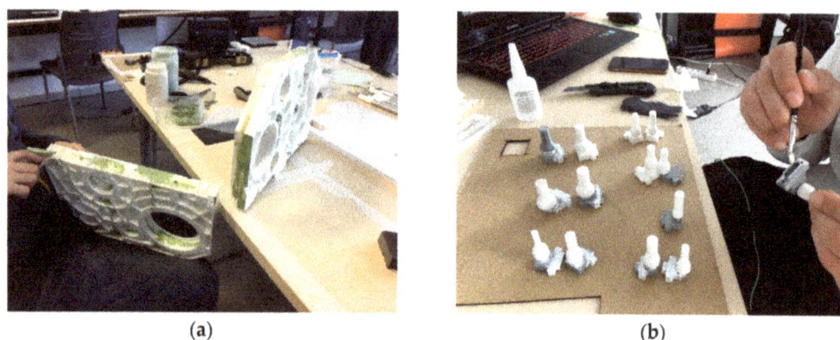

(a) **(b)**

Figure 11. (**a**) Process of joining two elements. (**b**) Process of matching color.

Finally, for its exhibition, a methacrylate urn is designed and manufactured, as well as a package to allow displacements without damaging the scale model. In this case, both elements have been made with cutting machines. In the case of the urn with laser cutting and in the packaging with CNC milling. Figure 12 shows the SO/PHI, from its digital model to its final manufacture stage as a HDM.

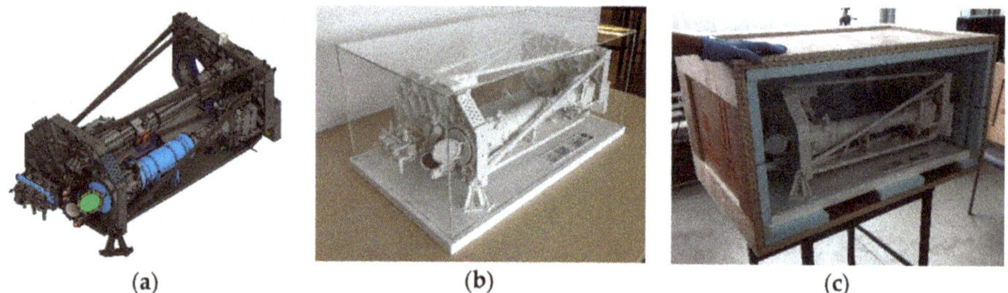

(a) **(b)** **(c)**

Figure 12. (**a**) SO/PHI digital model. (**b**) Final result of SO/PHI HDM. (**c**) HDM in its packaging.

4. Discussion and Conclusions

The case study consisted in the manufacture of a full-scale high-detail model (HDM) for its conservation as part of the aerospace heritage, through the low-cost digital manufacturing of the PHI instrument for the Solar Orbiter mission. This project was carried out under a collaborative approach, low cost manufacturing, and graphic engineering planning. For the collaboration of different teams and their supervision, techniques of organization and distribution of subassemblies were applied. This organization was essential to know at all times of the project what was the location of each element and its state of manufacture.

Aerospace engineering entails complexity that cannot be transmitted simply through texts, photographs, or simplified scale models. Although, these modalities serve for the conceptual preservation and the diffusion in educative environments of this heritage of engineering, the development of models of HDMs are necessary for the scientific and professional panorama. For this reason, it is proposed to add these scale models to the current categories [22].

Other authors have promoted digital graphic representation techniques such as augmented reality for the conservation of industrial heritage [15] or the application of BIM techniques for the representation of architectural heritage with a high level of detail [16], but the materialization of the work in a physical element can test and evaluate it [19]. This research generates a proposal that combines the level of detail in CAD modeling and benefits of its materialization.

This complexity has also been reflected in the need to solve problems through collaborative work and planning, with the particularity of working with geometric modeling in 3D. In this sense, results are presented that agree with Vila, Ugarte, Ríos, and Abellán [25], Fusion 360 has worked and allowed team work both synchronous and asynchronous, so that supervision tasks could be done through Fusion 360. In addition, printing times were optimized, as there were no downtimes between the low-cost digital manufacturing of highly complex engineering projects is possible, but far from being an instantaneous process. This experience fosters the generation of seven stages of graphic engineering and planning for the elaboration of models of high level of detail from the reception of a 3D digital model: identification and organization of subassemblies, analysis of the digital files, digital model repair, digital model restructuration to optimize their manufacture, adaptation of the models to the means of manufacturing, pre-allocation of the digital models to the manufacturing, and manufacturing and finishing.

The importance of HDM is supported by its role as a source of information in historical investigations [21]. However, the models have usually led to geometric simplifications due to the cost involved in representing the engineering project in detail [20,22]. This research contributes to the achievement of this high-detail model by the combination of two 3D printing types: fused deposition modeling and digital light processing (DLP). DLP have allowed to manufacture elements in photosensitive resin with very small dimensions at full-scale, being reproduced in the environment of the engineering project. This means a step forward in the detailed representation of engineering projects as opposed to architectural scale models that require geometric simplifications because they cannot be uniformly scaled [3].

This research has developed a full-scale high-detail model of the SO/PHI instrument for the Solar Orbiter mission of the European Space Agency through low-cost digital fabrication techniques and collaborative work. The HDM is presented as a tool that contains complex information that seems especially suitable for the aerospace sector, in which most of the designs are launched into space and there is no record beyond the digital model, texts, and photographs. In addition, seven engineering graphics procedures for the manufacture of HDMs have been provided to enable other professionals and scientists to contribute to the transfer and conservation of industrial heritage.

Author Contributions: Conceptualization, J.L.S. and V.L.C.; Methodology, software investigation and resources, J.d.l.T.-C. and M.D.D.-A.; Formal analysis and the writing of the manuscript, J.L.S. and V.L.C.; Supervision, project administration, and funding acquisition, M.D.D.-A. and J.d.l.T.-C.

Funding: This research was partially funded by Ministerio de Ciencia e Innovación, grant number HAR2017-85169-R: Fundicion artística de micro esculturas diseñadas por ordenador, mediante el desarrollo de técnicas de impresión 3D basadas en el procesado digital de luz". It has also been funded by National Institute of Aerospace Technology (INTA) and Andalusian Astrophysics Institute (IAA) through the project "Réplicas Dummies a escala 1:1 del instrumento espacial PHI para la misión Solar Orbiter de la ESA".

Conflicts of Interest: The authors declare no conflict of interest.

References

1. Instituto de Astrofísica de Andalucia: Unidad de Desarrollo de Instrumental y Tecnológico SOPHI. Available online: https://udit.iaa.csic.es/en/content/sophi (accessed on 24 May 2019).
2. Campbell, R.I.; Bourell, D.; Gibson, I. Additive manufacturing: Rapid prototyping comes of age. *Rapid Prototyp. J.* **2012**, *18*, 255–258. [CrossRef]
3. Stitic, A.; Robeller, C.; Weinand, Y. Experimental investigation of the influence of integral mechanical attachments on structural behaviour of timber folded surface structures. *Thin-Walled Struct.* **2018**, *122*, 314–328. [CrossRef]

4. Dimitrov, D.; Schreve, K.; de Beer, N. Advances in three-dimensional printing–state of the art and future perspectives. *Rapid Prototyp. J.* **2006**, *12*, 136–147. [CrossRef]

5. Lowke, D.; Dini, E.; Perrot, A.; Weger, D.; Gehlen, C.; Dillenburger, B. Particle-bed 3D printing in concrete construction—Possibilities and challenges. *Cem. Concr. Res. Pergamon* **2018**, *112*, 50–65. [CrossRef]

6. Stitic, A.; nguyen, A.; Rezaei Rad, A.; Weinand, Y. Numerical Simulation of the Semi-Rigid Behaviour of Integrally Attached Timber Folded Surface Structures. *Build. Multidiscip. Digit. Publ. Inst.* **2019**, *9*, 55. [CrossRef]

7. Oropallo, W.; Piegl, L.A. Ten challenges in 3D printing. *Eng. Comput.* **2016**, *32*, 135–148. [CrossRef]

8. Rojas-Sola, J.; Galan-Moral, B.; la Morena-De la Fuente, D. Agustín de Betancourt's Double-Acting Steam Engine: Geometric Modeling and Virtual Reconstruction. *Symmetry* **2018**, *10*, 351. [CrossRef]

9. Meier, C.; Saorín, J.L.; de la Torre-Cantero, J.; Díaz-Alemán, M.D. Alternative divulgation of the local sculptural heritage: Construction of paper toys and use of the minecraft video game. *Sustainability* **2018**, *10*, 4262. [CrossRef]

10. Saorín, J.L.; de la Torre-Cantero, J.; Meier, C.; Melián-Díaz, D.; Ruiz Castilló, C.; Bonnet de León, A. Creación, visualización e impresión 3D de colecciones online de modelos educativos tridimensionales con tecnologías de bajo coste. Caso práctico del patrimonio fósil marino de Canarias. *Educ. Knowl. Soc.* **2016**, *17*, 89–108. [CrossRef]

11. Acierno, M.; Cursi, S.; Simeone, D.; Fiorani, D. Architectural heritage knowledge modelling: An ontology-based framework for conservation process. *J. Cult. Herit.* **2017**, *24*, 124–133. [CrossRef]

12. Walsh, J.S.P. Protection of humanity's cultural and historic heritage in space. *Space Policy* **2012**, *28*, 234–243. [CrossRef]

13. Gorman, A. Culture on the Moon: Bodies in time and space. *Archaeol. J. World Archaeol. Congr.* **2016**, *12*, 110–128. [CrossRef]

14. Remondino, F. Heritage recording and 3D modeling with photogrammetry and 3D scanning. *Remote Sens.* **2011**, *3*, 1104–1138. [CrossRef]

15. Rojas-Sola, J.I.; Aguilera-García, Á. Virtual and augmented reality: Applications for the learning of technical historical heritage. *Comput. Appl. Eng. Educ.* **2018**, *26*, 1725–1733. [CrossRef]

16. Brusaporci, S. The Representation of Architectural Heritage in the Digital Age. In *Encyclopedia of Information Science and Technology*, 3nd ed.; Yoder, J., Henning, C., Eds.; IGI Global: Hershey, PA, USA, 2015; pp. 4195–4205.

17. De la Torre-Cantero, J.; Saorín, J.L.; Meier, C.; Melián-Díaz, D.; Drago-Díaz Alemán, M. Creación de réplicas de patrimonio escultórico mediante reconstrucción 3D e impresoras 3D de bajo coste para uso en entornos educativos. *Arte Individuo Y Soc.* **2015**, *27*, 427–444. [CrossRef]

18. NASA 3D Resources. Available online: https://nasa3d.arc.nasa.gov/models (accessed on 2 March 2019).

19. Hallgrimsson, B. *Prototyping and Model Making for Product Design*; Laurence King Publishing: London, UK, 2012.

20. Yongnian, Y.; Li, S.; Zhang, R.; Lin, F.; Wu, R.; Lu, Q.; Xiong, Z.; Wang, X. Rapid prototyping and manufacturing technology: Principle, representative technics, applications, and development trends. *Tsinghua Sci. Technol.* **2009**, *14*, 1–12.

21. Granado Castro, G.; Barrera Vera, J.A.; Aguilar Camacho, J. La maqueta de Cádiz de 1779. Utilidad militar o metáfora de poder. *Proy. Prog. Arquit.* **2016**, *15*, 16–29.

22. Salwa, S. Classifying Physical Models and Prototypes in the Design Process: A study on the Economical and Usability Impact of Adopting Models and Prototypes in the Design Process. In Proceedings of the DESIGN 2014 13th International Design Conference, Dubrovnik, Croatia, 19–22 May 2014; Marjanović, D., Štorga, M., Pavković, N., Bojčetić, N., Eds.; Faculty of Mechanical Engineering and Naval Architecture and The Design Society: Zagreb, Croatia, 2014; pp. 1–10.

23. Al-Samarraie, H.; Saeed, N. A systematic review of cloud computing tools for collaborative learning: Opportunities and challenges to the blended-learning environment. *Comput. Educ.* **2018**, *124*, 77–91. [CrossRef]

24. Autodesk Fusion 360: Fusion 360 for Hobbyist and Makers. Available online: https://www.autodesk.com/campaigns/fusion-360-for-hobbyists (accessed on 25 May 2019).

25. Vila, C.; Ugarte, D.; Ríos, J.; Abellán, J.V. Project-based collaborative engineering learning to develop Industry 4.0 skills within a PLM framework. *Procedia Manuf.* **2017**, *13*, 1269–1276. [CrossRef]

 *applied sciences*

Article

A Multi-Criteria Cataloging of the Immovable Items of Industrial Heritage of Andalusia

Juan Claver [1,*], Amabel García-Domínguez [1], Lorenzo Sevilla [2] and Miguel A. Sebastián [1]

[1] Department of Construction and Manufacturing Engineering, Universidad Nacional de Educación a Distancia (UNED), C/Juan del Rosal 12, 28040 Madrid, Spain; amabel.garcia@invi.uned.es (A.G.-D.); msebastian@ind.uned.es (M.A.S.)
[2] Department of Civil, Materials and Manufacturing Engineering. Universidad de Málaga (UMA), C/ Dr. Ortiz Ramos, 29071 Málaga, Spain; lsevilla@uma.es
* Correspondence: jclaver@ind.uned.es; Tel.: +34-91-398-60-88

Received: 31 October 2018; Accepted: 10 January 2019; Published: 14 January 2019

Abstract: Any research in any field needs an initial background, and in the same way, any decision should be supported by previous knowledge and study of the problem and its context. In the case of the industrial heritage, both the study of the typology and the decision making about the actions of conservation and reutilization of its assets must be based on a deep knowledge of the set of elements that the typology includes. All of that refers to the corresponding territory being analyzed, since the intensity and productive tradition will be different between each territory, region, or country. In that context, this paper represents the continuation of the main research line of the authors, and exposes their efforts to develop a useful tool for the study, management, and cultural promotion of the assets related to industrial heritage in Spain through the development of a multi-criteria catalogue of assets. Thus, based on the initial catalogue developed by some of the authors, this paper significantly increases the number of assets considered. In addition, it includes new classification criteria, reviews the observed trends, and establishes the future lines of work and suitable strategies for these kinds of initiatives.

Keywords: industrial heritage; cataloguing; methodology; multi-criteria; Andalusia; Spain

---

1. Introduction

In 2016, and as a part of a doctoral thesis, some of the authors elaborated a catalogue of immovable assets of the Spanish industrial heritage [1]. Despite the main objective of the research work being the design of a methodology based on multi-criteria decision-making tools, previous tasks that were identified included both the need of a wider catalogue of assets and a structure of classification criteria that applied to all of them. At that moment, the Initial Catalogue of the National Plan of Industrial Heritage [2], the Minimum Catalogue of the The International Committee for the Conservation of the Industrial Heritage (TICCIH)-Spain [3], and the catalogue developed by The DOCOMOMO (international committee for DOcumentation and COnservation of buildings, sites and neighborhoods of the MOdern MOvement) Foundation about the industrial buildings of the modern movement in Spain and Portugal [4], were the main cataloguing initiatives in the country. All of them were of great interest, especially with regard to their different approaches to the problem, but the number of assets that was considered in all of the cases was not large enough to be representative of this typology across the whole national territory.

The need to identify and classify the existing assets of value is considered in some of the most recognized documents about protection and actuation regarding the immovable assets of the cultural heritage, such as The Declaration of Amsterdam [5], which is focused on general architectural heritage, and The Niznhy Tagil Charter for the Industrial Heritage, which is specifically focused on this typology [6]. However, traditional cataloguing strategies must be revised and improved, both from a general approach to cultural heritage as well as from particular approaches to specific typologies [7,8], as the industrial heritage is [9].

Unfortunately, it is not difficult to identify examples of reuse actions above these kinds of assets in which the relation to the productive process has disappeared. These situations reveal that the concept of industrial heritage and the real value of these assets are not yet sufficiently clear. In that sense, some of the widely recognized definitions describe the industrial heritage as the result of the relation between a social model, the capitalism, and the productive resources of the mechanization [2,3]. This approach involves chronological limits that have been discarded by the authors of this work, who have considered both preindustrial and industrial assets within the catalogue. It is the singularity with regard to aspects such as the technological exclusivity or innovation, the layout and sizing of the productive spaces, the structure or the constructive techniques applied in order to response to the needs of the productive process, or the historical and social role that the activity had in the region, which are the kinds of features that can provide cultural value to these kind of assets [1].

Thus, in the case of industrial heritage, the recording of these assets should enable the understanding of their productive nature from different approaches [10]. In that context, a new catalogue was developed, including 1354 assets that were classified through 54 different criteria. This made it possible to propose a great amount of analyses by combining different criteria in each case. However, some improvement goals have been identified since then.

From a general approach to the initial version of the catalogue of assets, it is obvious that although the number of assets that was considered was significantly higher than the sets of elements, including those referred to in the previous catalogues, there are many more assets of interest that were not identified in that initial version. In addition, the national approach of that first experience required careful work regarding the identification and selection of assets, avoiding any kind of focus on particular territories in order to show real trends. Although this effort developed the analysis of the presence of the typology in a horizontal way through the whole country, it also made it more difficult to address a complete identification of assets. Thus, after the achievement of the initial goal of having a more useful set of assets and a criteria structure applied to all of them, the own needs of this new tool began to appear.

The review of the identification and selection process on the different regions was the first aspect that was identified as a new goal. As in the first attempt, the catalogue needed to be representative of the presence of the typology in Spain, so it is important to avoid any kind of extra effort in the identification of assets in any particular territory. Controlling this through a unique revision above the whole catalogue can be difficult. In addition, considering the number of assets included, reviewing the whole set is a tedious task itself. Therefore, the authors considered the development of subsequent studies focused on particular Autonomous Communities or provinces as the most reasonable strategy for this review process. Thus, some previous review works were developed in territories as the provinces of Vigo, Cádiz, and Huelva [11–13], and in the Autonomous Community of the region of Murcia [14]. All of these works enabled comparisons between the initially obtained trends and the new ones, and at the same time, during their development, some clues about new needs, approaches, and solutions were obtained.

In this context, the whole revision of the initial version of the developed catalogue will be the result of a set of partial review works focused on particular territories. In that sense, this work represents the first significant review experience for the catalogue, since it affects a particularly important territory: the Autonomous Community of Andalusia. This territory has some special characteristics, such as its high number of provinces, the variety of industrial sectors and their different presence through the eight provinces, or the high number of industrial assets identified in this territory in the initial version of the catalogue. All of these reasons make Andalusia one of the best territories for showing the potential of the catalogue as a work tool, and it was also the selected territory when the initial version of the catalogue was presented to researchers in this field [15]. This way, this new work is understood as a key previous step to reviewing the catalogue at the national level.

In addition to the classification criteria, the developed catalogue includes complementary information, such as web entries and geolocation information for each element listed. Thus, for all of the new assets that were added to the catalogue, geolocation information was included, too. This data field has a special value, because in some cases, the location of these assets is little known given the remote location of many of them. Industrial assets are often abandoned when productive activity ends. Sometimes, even the residential settlements that are the result of these productive activities disappear when production ends. Since the dissemination and promotion of this typology and its assets have a key role in achieving their protection, the geolocation information of a large amount of assets provided a great opportunity for promoting their visit [16]. However, the geolocation of these assets is also important in research contexts, such as for example the implementation of geographic information systems in the study of these elements [17,18]. Figure 1 shows a map generated using Google My Maps. Introducing in this application the data of the columns of the catalogue, which contains the name given to each asset and the geolocation information included for each asset, these kind of maps are automatically created.

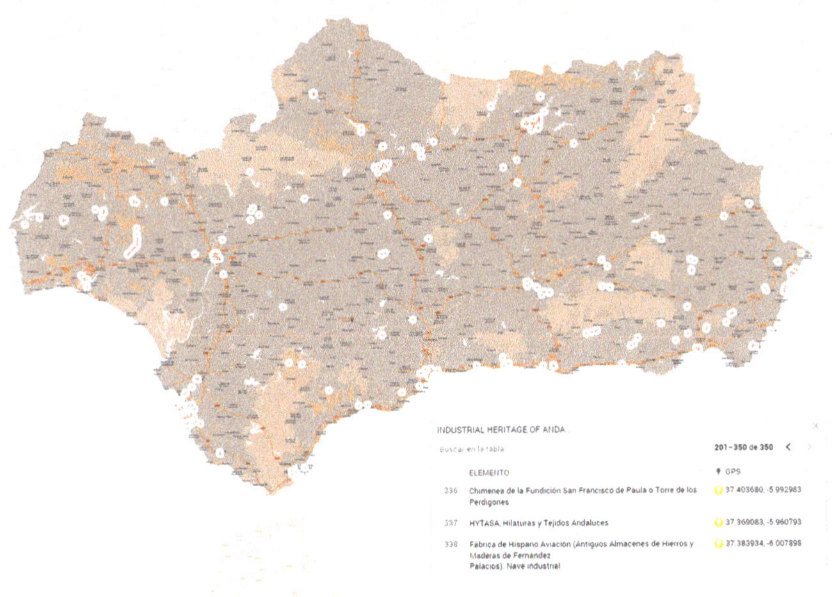

Figure 1. Visualization of the assets identified in Andalusia within a map created using Google My Maps and thanks to the geolocation information included for each asset in the catalogue.

Moreover, as it was previously introduced, the catalogue was born as support for the multi-criteria structures designed for the management and decision making about the reutilization of this type of assets, operating as one of the parts of a global methodology [19]. This methodology included three levels. Two of them were focused on the application of multi-criteria decision techniques, both for the valuation of the cultural interest of the analyzed assets and for the selection of the more compatible new activities. However, any decisions within these two levels had to be supported by a deep knowledge of this typology. In that context, a first version of a wide catalogue of industrial assets located in Spain was developed. In addition to the large number of assets considered in comparison to previous initiatives [2–4], the main characteristic of this catalogue was the application of a wide structure of classification criteria. This structure was applied to the whole set of assets, and by combining different classification criteria, it was possible to make a great variety of analyses of great interest in order to characterize this typology in Spain. Thus, the developed catalogue represents the necessary groundwork for those other criteria structures focused on multi-criteria decision-making, but these kinds of techniques are not applied within the catalogue.

During the operation of this methodology as a management tool, some new potential information fields of interest were identified. This work responses to these needs, and represents a first improvement of the classification criteria structure of the catalogue, increasing the number of criteria applied to the assets from 54 to 62. Multi-criteria approaches, such as the one developed by the authors, are considered of great interest, and it is actually possible to identify other proposals for the study of cultural heritage based on these techniques [20–22]. Thus, not only the previously obtained trends are reviewed, but also new possible analyses are exposed, and also in achieving this goal, the geolocation information had a key role, as will be exposed below.

Thus, this work proposes several analyses of the industrial heritage in the Autonomous Community of Andalusia, all of them by using the developed catalogue. These analyses enable characterizing this heritage typology in this territory and the review trends that were previously obtained, but they also show the potential of the catalogue as a tool for the study, promotion, and management of these assets. Furthermore, the incorporation of new classification criteria represents the first enlargement of the criteria structure of the catalogue. This means an increase in the variety of analyses and approaches enabled by this tool, and the improvement of the catalogue as the response to a need identified during its use as support for research activity [19]. In that sense, new possible lines of improvement are commented upon, and the main conclusions of the work are finally exposed.

2. Methodology

In this section, the main methodological aspects that were applied in this work are exposed. Thus, the growth process of the set of assets, the criteria structure of the catalogue, its use as analysis tool, and the special considerations of some of the classification criteria considered are defined and explained.

2.1. Set of Assets Increase

The first task addressed in this work was the identification and selection of new assets of interest in the territory considered as study case, which is in this case Andalusia. This searching work was not only an additive process focused on the increase of the set of assets considered, but also a review of that set and its capability of being representative of this typology in Andalusia. Perhaps the first goal that comes to mind when a recording initiative such as this one starts is achieving a wide enough set of elements to be representative of the reality that is to be characterized. In the case of the industrial heritage, the large number of assets falling under this term makes the identification of all of the elements difficult.

Nevertheless, the catalogue developed by the authors is not only a list of assets of interest, but also a whole criteria structure that is applied to all of them, so one of the main opportunities that the catalogue offers is the wide variety of analyses allowed by combination of these criteria. While accepting that including all of the existing assets of this type is very difficult, the wider the set of assets is that are considered, the more reliable and representative the trends of the territory that are observed through the run analyses. In this sense, the number of assets that is included in the catalogue for Andalusia was increased from 166 to 350, which means more than double the entries that were initially considered for this territory.

Figure 2 shows in blue the initial distribution of assets through the eight provinces of the Autonomous Community of Andalusia. The new assets that are included for each territory are marked in orange, and the total number of assets is indicated above each bar in the graph.

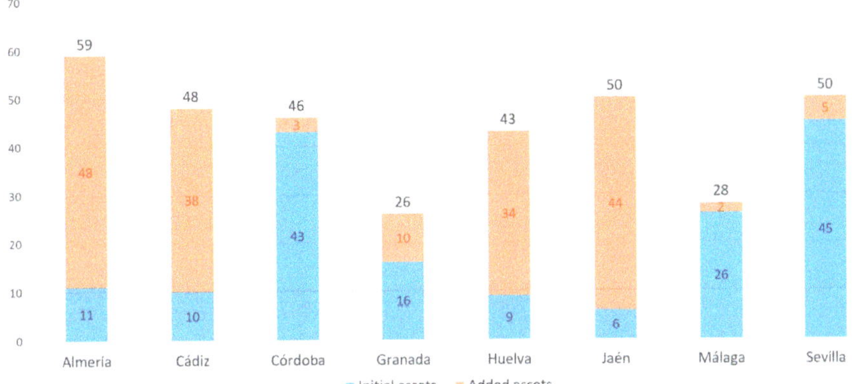

Figure 2. Assets initially considered for each province: new assets included, and actual number of assets identified.

Firstly, it is possible to appreciate significant changes from the initial distribution. Thus, territories with less importance at the initial distribution, such as Almería, Cádiz, Huelva, and Jaén, now have a greater role. On the other hand, territories with the highest number of assets at the initial distribution—Córdoba and Sevilla—and the ones in intermediate positions—Granada and Málaga—have experimented much lower increases. This can be perceived as an attempt to equilibrate the distribution trend, but similar resources and efforts were inverted in all of the cases. The new territorial distribution after the actualization of the catalogue will be analyzed in more depth in the later sections.

Another consequence of this study focused on Andalusia is the imbalances that the actualization of the sample in this territory causes in the territorial distribution at the national level. Before this work, Andalusia was the third territory by the number of assets included in the catalogue, only behind Catalonia and Madrid, and very close to the Basque Country. Actually, its 350 assets make it the leading territory in assets contribution to the catalogue. Due to this, similar works must be developed for each one of the other autonomous communities, and for those future works, the present study must establish a reference baseline.

2.2. Criteria Structure

In this section, the classification criteria structure of the catalogue is briefly exposed, differentiating between initial and new criteria, and showing the logic and the functioning of the catalogue. However, after the general overview of the criteria structure, some guidelines for the application of some classification criteria are commented. Despite all of the criteria considered being attributes that are easy to understand, their application is not always so simple. Classifying any kind of elements in relation to a particular aspect and under a finite number of categories can cause problems. When such a wide set of elements must be processed, there will be particular situations in which the boundaries between the classification categories can be confusing. Having some simple selection guidelines will be very useful in those cases.

Figure 3 shows the criteria structure of the catalogue. As shown in the figure, the criteria that are considered can be grouped along three general categories. The first one includes information that is able to define each asset from different approaches. Thus, the proposed designation for the asset, information concerning the location of the asset, the definition of the scale of the asset, the technological level of the asset, the role in the productive process, the corresponding industry sector, and an interactive link to a website with content related to each particular asset are included in this first category of the catalogue. A second group of criteria is focused on the presence or not of each asset into cataloguing initiatives of special interest, as well as the protection, when it exists, as a Good of Cultural Interest (BIC). Finally, a third group of asset enables identifying those assets that are reused for new activities, and for those cases, the nature of these activities also.

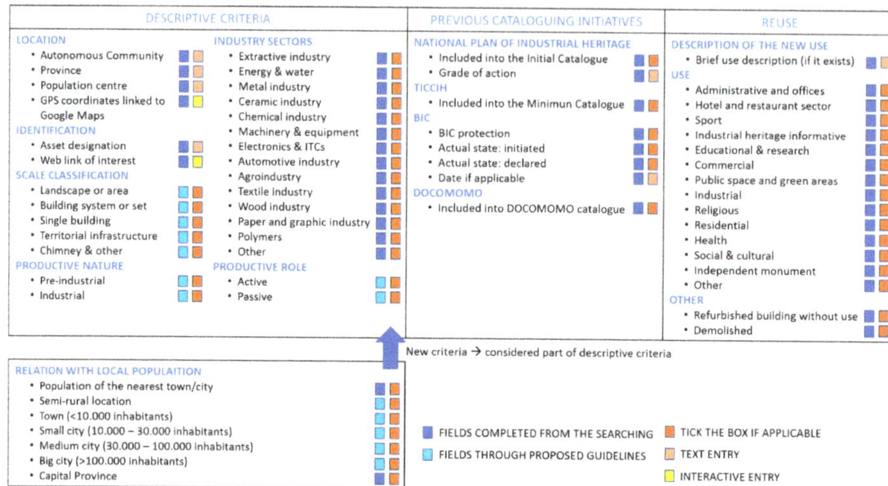

Figure 3. Criteria structure of the developed catalogue: information obtainment process and type of information.

Figure 3 also indicates the nature of the information contained in each field. First, with regard to the information obtainment process, it is necessary to differentiate between those fields completed through only information searching work and those which required some interpretation guidelines. Secondly, the way in which the information of the different fields is shown in the catalogue is also indicated. The most common procedure is a simple mark in the corresponding field when that characteristic is noticeable in the analyzed asset. In other cases, text entries are necessary in order to describe the information of that field. In addition, two fields with interactive information are included for each asset. One of them is linked to online information of interest for each single asset, and the other one contains the global positioning satellite (GPS) coordinates of the corresponding asset, and is linked to Google Maps.

New Criteria Added to the Structure of the Catalogue

In addition to the criteria considered in the initial version of the catalogue, new criteria were incorporated to the structure. This new set of features can be understood as new descriptive criteria, since they contribute to the characterization of the analyzed assets. Figure 3 suggests this idea. The aim of these new criteria is to determine the relation of the industrial assets considered with the existing local population. This relation, and the potential synergies issuing from it, are of great interest when the possible reuse of one asset is being discussed. In fact, as previously mentioned, the interest of including this group of criteria was detected through the use of the catalogue as part of a global methodology, in which the information contained in the catalogue supports multi-criteria decision structures for the selection of the most appropriate new uses for these kinds of goods [19]. As in the case of the previously exposed criteria groups, the nature of the information contained in each field is also indicated in Figure 3.

Table 1 illustrates how the eight criteria added to the structure of the catalogue try to define the relation of the industrial assets that were analyzed with the local population. Thus, two implicit classification levels can be identified through these eight criteria. First, they allow differentiating between rural and urban environments. Then, inside each of the two main categories, new particular situations are considered. For rural environments, the relationship with the local population is then analyzed. In the case of assets located in urban environments, the size of the population center is indicated. Special aspects are also considered, such as the special role that the province capitals have in each region, and the number of inhabitants of the nearest population center, according to the data provided by the National Statistical Institute of Spain [23].

Table 1. Approach of the new classification criteria added to the catalogue structure.

First Level of Classification	Second Level of Classification	Catalogue Denomination
Rural environment	Relation or not to any human settlement	Rural location Semi-rural location Town
Urban environment	City size	Small city Medium city Big city
Complementary information	Special role Population measurement	Province capital Population of the nearest town

2.3. Application Guidelines

As exposed before and indicated in Figure 3, some classification criteria need basic guidelines when they are applied to the set of assets of the catalogue. In this section, some orientations are proposed in order to make the classification through these criteria easier and clearer.

2.3.1. Scale

One of the most interesting characteristics of the industrial heritage, as the typology of the cultural heritage, is the variety of scales in which these assets can appear. The criteria structure of the catalogue includes five different categories for the classification by scale: chimneys and other, immovable asset, system or building set, landscape or area, and territorial infrastructure. All of these categories are easy to understand through examples, but also it is possible to give many examples in which the boundaries between categories are not so obvious. In order to help with classifying each good by this criterion, the authors defined a set of aspects that can be perceived in a different way from the perspective of each classification category [1]. Figure 4 resumes the initial guidelines.

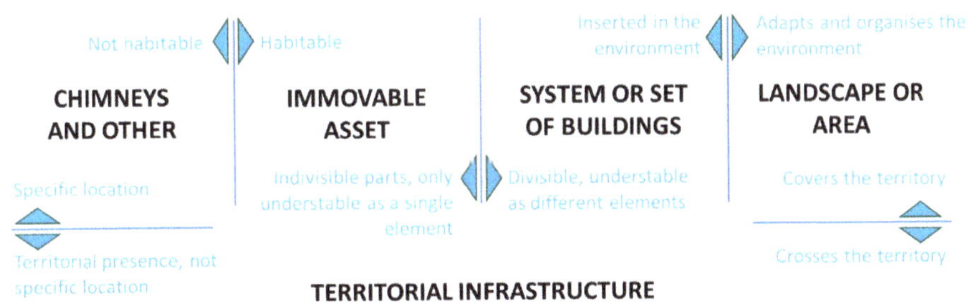

Figure 4. Graphical guidelines for the classification through the scale criterion.

The category "chimney and other" is used when only the remains of some isolated parts form the original state of the asset. The most usual example of this situation is industrial chimneys, but other elements such as melting furnaces are also included in this group. These kinds of elements are not movable assets, but they are not immovable assets, either. Nevertheless, some of these elements can be really large, as in the case of blast furnaces, so they can even perceived as buildings by many people. Figure 4 identifies the habitability as the key aspect to be considered when the studied good did not belong clearly to either of these two categories. Chimneys, furnaces, and other similar elements have been built using the same constructive systems used in the industrial buildings in which they were included, such as for example using bricks, and they can be big, too; so, effectively they can be really close to the building scale. However, they were not designed to be inhabited, and there are not real living spaces in them.

The limits between the category of "immovable asset" and "system" or "set of building" will be confusing. It is easy to distinguish between a single building and a set of them. However, there are buildings that are the result of the sum of different spaces that define various independent volumes. In these cases, it seems as if various different buildings had been joined into only one. This idea can be exposed clearer in the opposite sense. It is usual that industrial facilities, with a long enough period of activity, experiment with changes due to the adaptation of the installations to new needs. One often action in that context consists in the covering of the space between two industrial buildings in order to expand the facilities. However, this does not mean that those two separate buildings are only one, even though they can be seen in that way. When classifying an asset in that situation, its analysis will not be focused only in the constructive appearance as a single building, and its more complex nature will be detected.

Thus, the same approach can be applied to other similar situations. There are industrial buildings in which it is possible to identify enough independent elements. However, the application of this idea can be confusing, so a clear guideline must be established. An approach to the problem through reuse actions in large industrial facilities can provide the key aspect to be considered. When this type of asset is recovered for new uses, usually the whole facility is not reused, but some elements are, even for different activities, and this is a demonstration of their independence. Thus, when analyzing great buildings, which also could be understood as the result of a sum of different volumes and productive spaces, thinking about the real possibility of acting on only some of those above parts is what can help determine whether enough independence exists between them for them to be considered as different elements, or not. Figure 5 illustrates this idea graphically: on the left and right are representations for the concepts of a single asset and a set of assets, and in the middle area of the figure, intermediate situations are shown. Thus, the capability or not of acting only on particular parts is what determinates the classification as a single asset or as a set of assets.

2.3.2. Technological Level

Two classification fields are included regarding to this aspect. Thus, assets can be identified as "pre-industrial" or "industrial". This classification can seem easy, but when you classify an element you are applying some criteria, and sometimes the decision about which are the appropriate criteria can be complicated and the subsequent application of the selected criteria can also be confusing.

The definition of industrial heritage that was proposed by the National Plan of Industrial Heritage of Spain in 2001 [2] identified the assets that compound this typology as the one result of the relation between a social model, capitalism, and the technological resources of the machining, which constrains these assets within the period between the second half of the 18th century and the last third of the 20th century. This temporal limitation is similar to the one proposed by The International Committee for the Conservation of the Industrial Heritage (TICCIH), considering the same start and an as of yet undefined end that goes on until today.

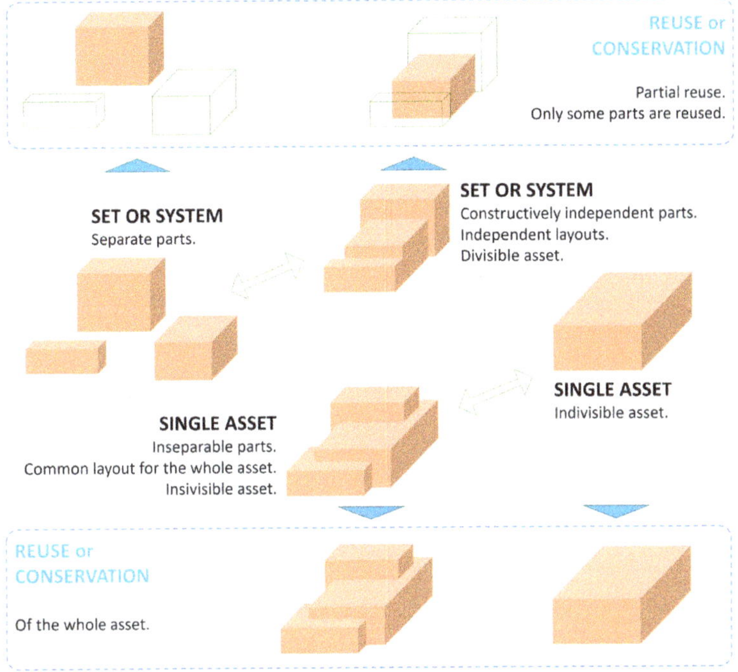

Figure 5. Graphical guidelines for the classification as a single asset or as a set of assets.

These general approaches define a limit between an industrial era and a pre-industrial era, and that idea was discarded in this work. The authors understand that the concept of industrial heritage must be able to cover and include any element—both movable and immovable assets—that are capable of helping the understanding of the operation of productive processes of interest, and also the life around these activities. The corresponding historical period is important information and in this study it is used as a classification criterion, but it is not a reason for the exclusion of a group of assets of interest, nor for the preference of some of them.

Thus, after the decision includes both possibilities in the catalogue, the guidelines for making the decision in each particular case must be established. The limits between industrial and pre-industrial assets might sound simple, but in many cases, it will not be so clear to decide in which group a particular asset must be included. The operation period of the asset is not a key aspect in this respect. Pre-industrial assets such as windmills were operating (and also were built) during the period defined by the National Plan and TICCIH as industrial. So, what criteria could be of help?

For that purpose, the authors established two main aspects that must be analyzed before the classification of each particular asset as pre-industrial or industrial [1]. On the one hand, there is the technological level of the asset, but this level is from a particular point of view that is focused on the power source that was employed. When this source is linked to the natural resources of the environment, the location of the asset is conditioned by the need of proximity to that resource, and this is understood as a pre-industrial attribute, in contrast to industrial situations in which steam engine or electric energy are the usual power sources. However, special situations can exist, so a second aspect is analyzed: the productive level or capacity. When the productive activity supplies the area near the asset, the productive level is understood as local, which means a pre-industrial attribute for the classification into the catalogue. When the asset is focused on exporting the products to other areas or regions, the productive volume is higher, and this is understood as an industrial attribute.

By combining the conclusions of both analyses, it is possible to select the industrial or pre-industrial nature of each particular asset. There will be situations in which these conclusions will point in opposite directions, and then the trend that is valued as the one of higher relevance will be one considered for the final classification of the asset.

2.3.3. Relation with Local Population

As exposed before, one of the main functions of the developed catalogue is the support of other initiatives of study, promotion, or actuation. In that sense, the catalogue was initially integrated in a global methodology for the study and management of the industrial immovable assets that have been developed by some of the authors, concretely for the selection of the most compatible new uses for particular assets, in order to minimize as much as possible the impact of the adaptation to the new activity [19]. Precisely in that context, the interest of including some new criteria in the catalogue was identified. When one of these assets is recovered for a new activity, one of the most important aspects is the success of this activity. When the new use fails, two new scenarios emerge for the asset. On the one hand it can be again in a situation of abandonment for some time, which probably will deteriorate the asset in some way. On the other hand, it can be adapted again for a different new use, and that process will have some kind of new impact. Both situations are undesired, so making the right selection of the new use is a key aspect.

In that sense, the criteria structures of the decisions of the mentioned global methodology demand the inclusion of new criteria into the developed catalogue. The environment of the assets is very important for the success of the potential new uses. For example, some uses will benefit from highly populated environments, while other activities can be more appropriate for rural contexts. Thus, in addition to the location criteria that are initially included into the catalogue, criteria related to the rural or urban context of each asset considered, and the size of the surrounding villages and cities have been included in the catalogue. Thus, eight new categories were considered and included into the catalogue as new classification criteria. These new criteria were listed below in Table 1, which also shows the logic behind the new criteria incorporated into the catalogue, but not the needed application guidelines.

In the case of rural environments, two possible situations are considered. The classification as a "rural location" is used when the analyzed asset has no direct relation to any nearby village or town, either because a nearby village or town does not exist, or because there is not a clear connection with the analyzed asset, for example a road. Thus, this criterion of classification shows a certain grade of territorial isolation. The criterion called "semi-rural location" can represent assets out of the urban fabric, but with certain connection and relation to some nearby town or village, for example due to its close proximity or for the existence of a road, and also it can be used for assets associated to small groups of houses without their own real administrative or fabric structure. As it is possible to appreciate, assets classified with this criterion will be neither included in large urban structures nor connected to them, but they will not have the previous level of territorial isolation. If potential synergies with human settlements can be identified, then the potential demand of certain new uses can be taken into account.

On the other hand, four categories are defined for assets located in urban environments. The category called "town" is used for populations of less than 10,000 inhabitants. When the number of inhabitants is higher than 10,000 the term city is used, while "small city" is when the population is between 10,000–30,000 inhabitants, "medium city" is when the population is between 30,000–100,000 inhabitants, and "big city" is when the population is higher than 100,000 inhabitants. Figure 6 shows graphically the key aspects for decision making about the most appropriate category of classification in each case when these new sets of criteria are applied to each particular asset.

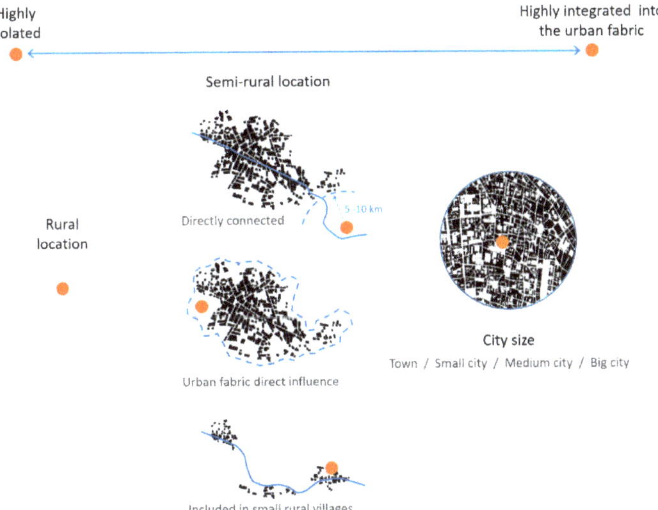

Figure 6. Guidelines for assets classification by criteria related to the grade of relation with human settlements.

In addition, the special role as "province capital" is also considered. When the population intervals were defined, the number of inhabitants of the main cities and towns of Andalusia was consulted in the database of the National Institute of Statistical of Spain [23]. Table 2 shows that, together with the eight province capitals, other cities have an important population in Andalusia, cities in which different industries are present, and contribute with some elements to the catalogue, so this additional classification criterion was considered of interest.

Table 2. Cities with the highest population in Andalusia. Source: National Institute of Statistical of Spain [23].

	Population		
	2015	**2016**	**2017**
Sevilla	693,878	690,566	689,434
Málaga	569,130	569,009	569,002
Córdoba	327,362	326,609	325,916
Granada	235,800	234,758	232,770
Jeréz de la Frontera (Cádiz)	212,876	212,830	212,915
Almería	194,203	194,515	195,389
Huelva	146,318	145,468	145,115
Marbella (Málaga)	139,537	140,744	141,172
Dos Hermanas (Sevilla)	131,317	131,855	132,551
Algeciras (Cádiz)	118,920	120,601	121,133
Cádiz	120,468	118,919	118,048
Jaén	115,395	114,658	114,238

The application of this group of criteria, as well as the other ones included in the catalogue, is supported by the geolocation information included for every asset of the catalogue. This aspect is exposed in the following section.

2.3.4. Geolocation Information as Support for Cataloguing Criteria

Geolocation information represents a data field with special potential from many different points of view. As mentioned before, for each asset included in the catalogue, the GPS coordinates are shown, and when selecting this field for a particular asset, the location of that element is shown in Google Maps. Having every asset of the catalogue linked to Google Maps [23] has an enormous potential and offers many opportunities, both as a promotion tool for this typology and as support information when particular assets must be classified through some of the criteria considered in the catalogue.

Figure 7 shows some of the main possibilities that this data field offers. Having this information for all of the assets of the catalogue, it is possible to make maps of assets by using Google My Maps. In these maps, it is possible to include a set of assets that has specific characteristics, such as the location in a particular region, the activity sector, or any other aspect included in the catalogue as one of the classification criteria. In addition, for each single asset, it is possible to make a virtual approximation or visit through the functions included in Google Maps [24] such as aerial views, Street View, and the photographs loaded by users.

Figure 7. Functions included in Google Maps.

Thus, this resource has significant potential applications as a tool for the diffusion and promotion of these assets, as well as for their visitors, so the interest in the field of tourism is considered relevant. However, in addition, during the application of the classification criteria considered in the catalogue the geolocation information also has a key role. In many situations this information will help to apply properly some classification criteria, mainly the scale criterion and the relation with local population. In both cases the geolocation information and the different virtual view modes allowed by it, make possible to evaluate and select the most appropriate category for each analyzed asset.

3. Results and Discussion

Figure 8 allows a combined interpretation of the distribution of assets, both by their location into the different eight provinces of Andalusia, and by the sector of activity. Thus, a vertical reading of the graph provides an interpretation of the distribution of assets for each sector of activity through the eight provinces. When the graph is read horizontally, the distribution of assets into the different sectors can be observed for each territory. In addition, at the top of the graph and on the right side, the general distributions of the total number of assets that have been identified for each territory and for each activity sector are shown. Thus, regions and activities with special relevance and presence within the set of assets that are being considered can be identified, and then distribution trends can be observed through vertical and horizontal interpretations of the graph.

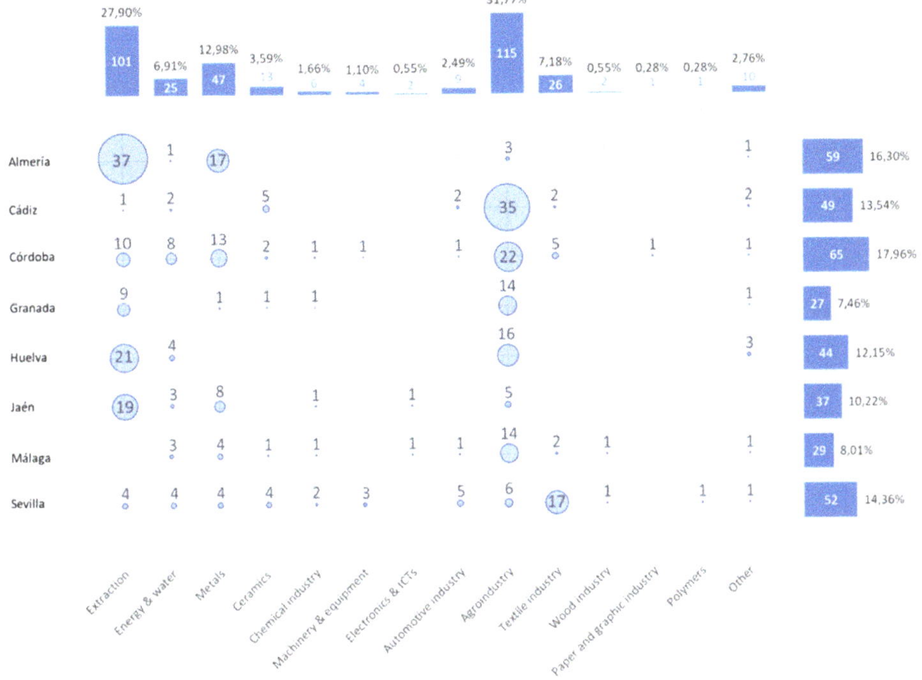

Figure 8. Distribution of assets simultaneously considering location and sector criteria.

At the same time, the different scales of the nodes in the graph enable identifying situations of special relevance. Examples of this are the high presence of assets linked to extraction activities in Almería and agroalimentary industries in Cádiz. The identification of these high presences is of high interest, and will derive in deeper studies in order to appreciate the circumstances behind these trends. Thus, in the case of Cádiz, by considering the scale criteria, it is possible to appreciate the significant presence of an immovable asset scale. When the set of assets is consulted, it is possible to identify a great number of windmills, which are the elements that make the presence of this activity sector so important within the distribution trends that are analyzed. Figure 9 shows this situation graphically. In addition, photographs of some examples of the identified windmills are included.

From a different point of view, the reutilization of this kind of assets for new activities is another aspect that is of the highest importance. Understanding the developed catalogue as a support tool for hierarchical multi-criteria structures, the identification and then the analysis of the reutilized assets represent key information.

In Figure 10, using a graph similar to the one used in Figure 8, the distribution of assets that simultaneously considers the criteria related to the location and the new uses is analyzed. At the top of the graph, the distribution of assets that is reutilized for each possible new use is shown, considering the assets located in every province. Thus, this trend represents the global distribution for this approach for the Autonomous Community of Andalusia. This analysis is of high interest, so it enables identifying the uses that have been able to activate new possibilities for assets that had lost their original function, and which needed a new one for their sustainable conservation. Consolidated examples of assets reused for new activities represent information of great value in order to establish appropriate strategies for the reutilization of this type of assets, providing examples of both good and bad practices.

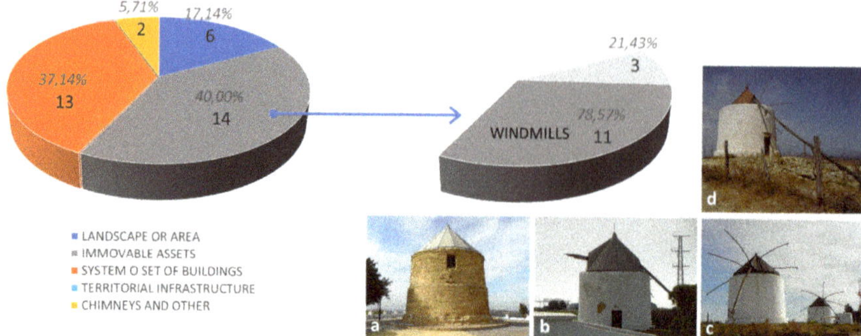

Figure 9. Distribution of assets for the scale criterion applied to the agroindustry assets identified in the Province of Cádiz. The presence of windmills and some examples in Vejer de la Frontera: (**a**) and (**b**) windmills at "Loma de la Buenavista"; (**c**) windmills that neighbor San José; (**d**) windmill "la Cruz de Conil".

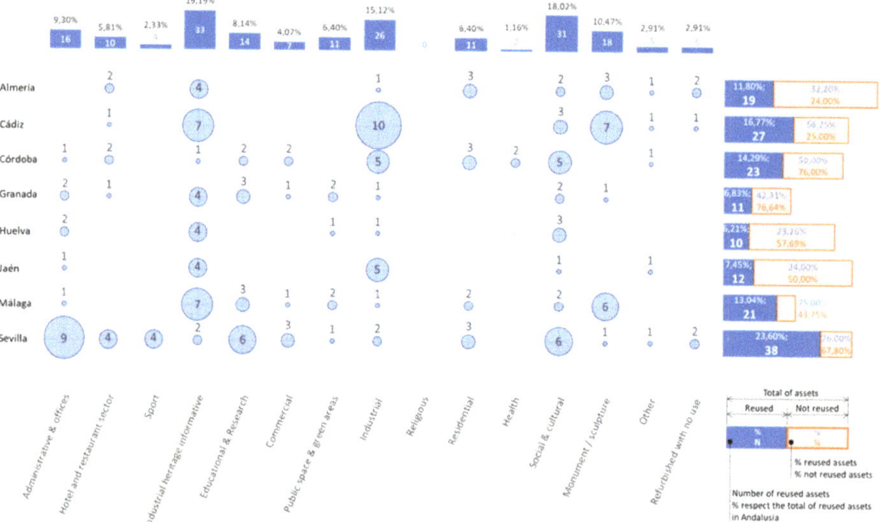

Figure 10. Distribution of assets simultaneously considering location and the new use when it exists.

In addition, it is possible to make a similar horizontal interpretation of the graph for each province. This enables appreciating the particular trends within each territory and identifying the most relevant new uses in each case.

When Figure 10 is interpreted vertically, it is possible to appreciate the presence of the assets identified for each new use within every particular territory. This enables identifying the special relevance of some sectors of activity in particular territories. For example, in the case of the use identified as "administrative and office", the graph shows the especially significant importance of the Province of Sevilla with regard to this use. At the same time, for the uses with highest presence, which are the ones identified as "industrial heritage informative" and "social and cultural", the graph shows trends with more balanced presence distributions in each territory, but it is also possible to identify the most significant provinces in each case.

On the right side of the graph, the total reutilization of assets observed in each territory is shown. This global analysis does not take into account the specific nature of the new activities, but rather considers the total number of assets reused through the different new uses in each province. Both the number of assets reused in each territory and the percentage of them within the total number of reused assets in Andalusia are indicated. Furthermore, the amount of reused assets in each territory is related to the total number of assets considered in the catalogue for the corresponding province. Thus, for each territory, the blue bar corresponds to the reused assets, and the orange bar corresponds to the ones that have not been recovered for new activities. Within each orange bar in the graph, the percentage of reused and not reused assets for each territory is indicated, using blue and orange text respectively in order to add clarity.

Furthermore, as in the graph shown in Figure 8, the size of the nodes makes a fast and easy identification of special and relevant situations possible, such as the previously mentioned presence of the use that is denominated as "administrative and offices" in the Province of Sevilla, or the significant number of assets dedicated to industrial activities in the Province of Cádiz. The identification of these and other high presence situations within the graph enable proposing and defining new analyses when necessary.

Since information about the reutilization of these assets for new activities is important, both as a sustainable strategy for their conservation and as support for the multi-criteria decision structures included in the global methodology developed by the authors [1,19], further analyses are exposed in order to show the potential of the catalogue as a useful and flexible tool.

Figures 11 and 12 propose new approaches by considering simultaneously three different categories of criteria: location, new uses, and scale. Figure 10 identifies "industrial informative heritage" and "social and cultural" as the uses with higher number of assets. On a second level, it is possible to identify three uses: "administrative and offices", "educational and research", and "monument/sculpture". These uses of special presence represent the most consolidated reuse strategies of this typology in Andalusia, and analyzing the possible influence of characteristics such as the scale of the reused assets represents a very interesting approach. Thus, Figures 11 and 12 incorporate a new criterion into the analysis shown in Figure 10, the scale of the corresponding assets, in order to identify possible relations between the activities established and the scale of the affected assets.

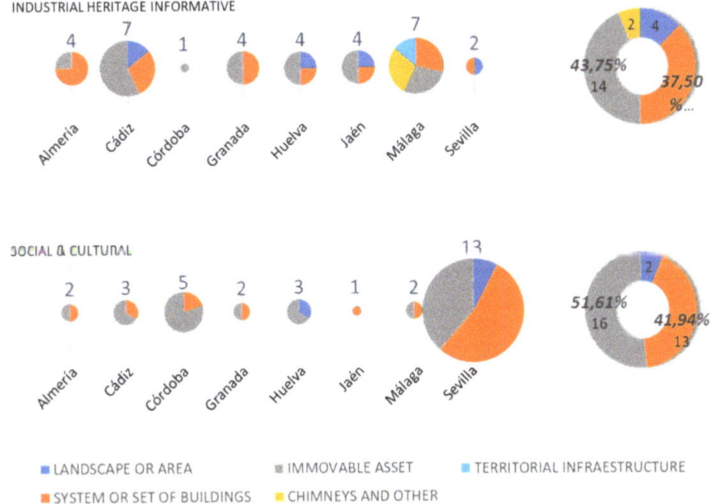

Figure 11. Distribution of assets within the two new uses with the highest number of assets and also considering the location and the scale criteria.

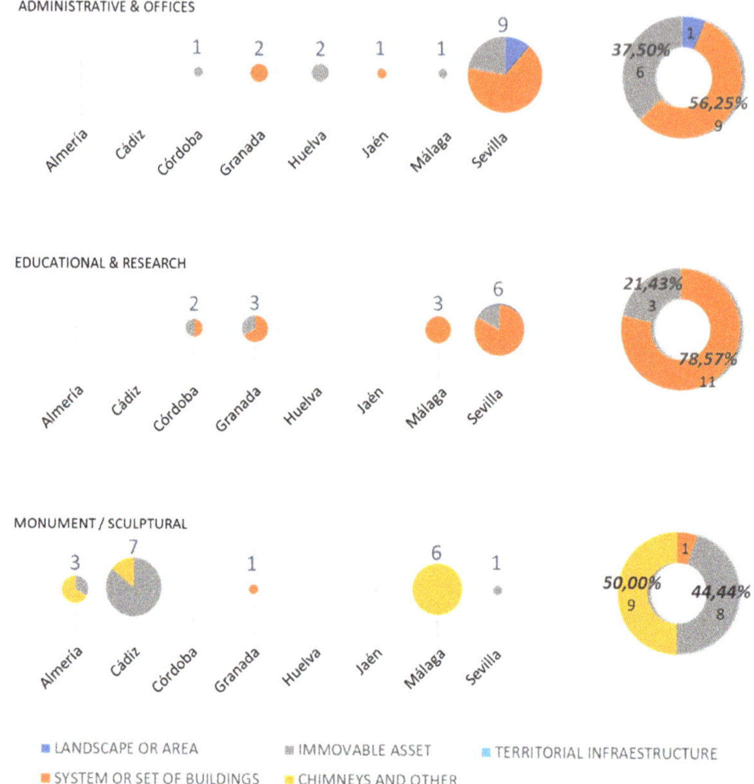

Figure 12. Distribution of assets within the new uses with a significant number of assets and considering the location and the scale criteria.

At the top of Figure 11, the distribution of assets through the eight provinces of Andalusia and the presence in each case of the scale categories considered into the catalogue are shown for the use "industrial heritage informative". At the bottom of Figure 11, the same approach is applied to the assets corresponding to the use "social and cultural". In both cases, a representation of the total distribution of scales for the analyzed use is shown on the right side of the figure. The same approach is applied in Figure 12 to the other three analyzed uses.

Observing both figures, it is possible to appreciate some trends of interest. Firstly, it can be noticed that the "territorial infrastructure" scale does not have presence. This highlights the greater difficulties that the assets of this scale present in terms of reuse for new activities. Regarding the use of "industrial heritage information", the other four scale criteria that are considered in the catalogue have representation, and in most of the cases, in several different territories. This heterogeneity is due to the nature and the objectives of this possible use, whose main purpose is not only the conservation, but also the dissemination and promotion of the concerned asset and its heritage values. Thus, it is not so much the scale demanded by the new use, but the scale of the original asset itself that persists, since the scale itself is a distinctive feature, and a characteristic of the value that must be preserved and shown. In the other cases, it is possible to identify a clear prevalence of some scales over others.

In the case of the use named "social and cultural", two scales have a clear predominance without signs of a significant preference for either one of them. Thus, both the "immovable asset" and the "system or set of buildings" scales emerge as the main alternatives for this use. In addition, two examples of "landscape or area" scale appear. This is due to isolated situations derived from the classification guidelines established for the construction of the catalogue [1]. Industrial landscapes and areas represent dimensions clearly different from other scales such as industrial facilities composed by a set of buildings, so this kind of elements must be identified as such. However, other criteria of the classification structure applied into the catalogue may affect only some of the recognizable single elements that are included within that landscape or area. The reuse of some of the elements while the others are abandoned, or the legal protection only of some of them, are examples of this situation. This way, the general guideline applied in the catalogue is the identification of the landscape or area scale as an asset into the catalogue, and then the identification of those elements that have special classification circumstances through the classification criteria structure as new and different assets. Due to this, while each one of the elements that are included within that greater scale asset usually present a single new use (in cases in which they have been recovered for a new activity), the corresponding asset of the greater scale will consider the uses identified with all of them, although only one of them is in fact the use of that landscape or area. Thus, the presence of assets with a scale of landscape or area for social and cultural use is a consequence of the limitations of the classification criteria structure, not a real relation between that scale and that use.

In the case of "administrative and offices" use, a certain preference for the scale named "system or set of buildings" is shown, but a clear trend is not able to point out the relation between the needs of the use and the characteristics of the scale criterion. Furthermore, regarding "educational and research" use, the preference for this scale is very significant. This is because in this case, the relation between the needs of the use and the distinctive characteristics of each scale exists, since a main need of educational centers is playgrounds, as spaces that can be disposed between the buildings and elements of some typical industrial facilities layouts. These kinds of conclusions are of great value when multi-criteria decision structures for the selection of the most compatible new uses are built [1,19].

Finally, the third graph included in Figure 12 shows the scales linked to "monument or sculpture" use. This use is mainly focused on chimneys and other elements, such as furnaces, which are inhabitable due to their scale. Thus, this kind of asset does not have the capacity to accommodate any kind of new uses. Nevertheless, they are industrial elements of interest, so they must be identified, classified, and protected. In Figure 12, it is possible to appreciate the significant presence of this scale, but also the "immovable asset" scale has presence, especially in the province of Cádiz. This is due to the significant presence of windmills in this territory, which was previously exposed and also graphically illustrated in Figure 9.

New criteria added to the classification criteria structure of the catalogue can be also analyzed, both from approaches focused on their own information and combined with other criteria. As exposed before in the section "relation with [the] local population", these criteria try to show the possible synergies between each asset and the nearest population centers. Two main aspects affect this relation: the population of the corresponding village, town, or city, and the road communication and general accessibility to the asset from the corresponding population center. Both aspects are important when new uses are proposed, but a connection between both elements is the key issue. A large population provides potential users with the new activities, but if there is no real relation between the asset and the population center, its size will not be relevant. This idea is graphically explained in Figure 6. Thus, six relation criteria categories are defined, which can be understood as seven when the special role of "province capital" exists. Firstly, Figure 13 analyses the size of the nearest population centers for each asset of the catalogue.

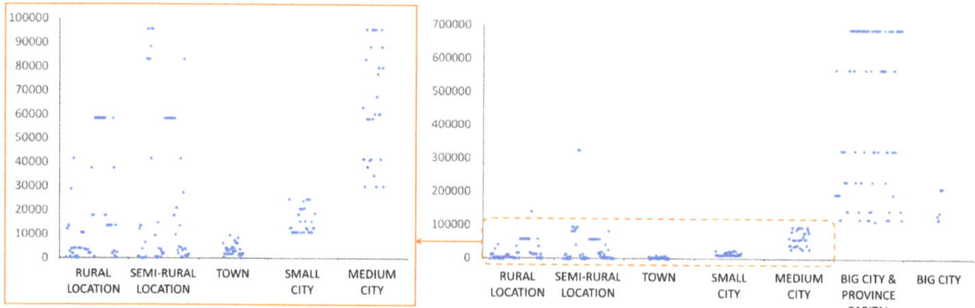

Figure 13. Distribution of assets within classification categories focused on the relationships with population centers by considering the number of inhabitants.

As exposed in Figure 6, three main situations can be distinguished by considering the relation between the asset and the nearest center of population; rural locations, semi-rural locations, and city or urban locations. Within urban contexts, four subcategories are defined (town, small city, medium city, and big city), making it possible to distinguish different sizes. Size is a key aspect to understand how a city works. Road and communication systems, infrastructures, and services are examples of aspects with different dimensions and complexity levels according to the type of city. However, rural and semi-rural locations have no relation to the size of the nearest population centers. In that sense, Figure 13 analyses the different sizes identified for the assets classified within each one of the categories that were considered.

It is possible to appreciate how population centers that are linked to assets with rural and semi-rural locations have in many cases significant sizes, which are similar to the ones identified for small and medium cities. On the left side of Figure 13, a zoom of that part of the graph is made in order to appreciate these situations more easily. The real potential synergies with population centers when a possible new activity is evaluated for an asset will be conditioned by aspects such as the size of those villages, towns, and cities, so these approaches are of interest when reutilization studies are being developed. On the other hand, Figure 13 reveals how most of the assets that have been identified for the category of big cities are located in province capitals, although there are a few located in some of the other big cities identified in Table 2. Despite the size of the cities, their social and administrative roles are also important, so not only the size is relevant, but also being the capital of a region has a great value, and can make it easier to develop these kinds of reuse projects.

In that sense, Figure 14 tries to identify the relations between these location categories and the presence of reuse initiatives. As it is shown, there is a great leap in this aspect between rural and semi-rural locations and the rest of the categories, which are associated with urban contexts. This highlights the influence of context circumstances for the success of reuse actions, but the identification of those examples that exists in rural and semi-rural environments and the strategies applied within them are also interesting.

Within the urban categories, the highest reuse values are identified for bigger city sizes. Thus, medium and large cities are the categories with a greater percentage of assets reused for new activities. In the case of big cities, the role as a province capital is able to increase this percentage, but not in a really significant way. It is also interesting that medium-size cities obtain a major reuse percentage. This can be due to the more aggressive urban growth processes that usually happen in big cities and which significantly modify the uses and the land value, which are aspects that make the conservation of old industrial facilities very complicated from a business viewpoint. In addition, the bigger the city is, the more difficult the feelings of belonging to the social history and tradition of the place among the population. These feelings are able to contribute to the conservation of the elements of this typology in a key way, or not.

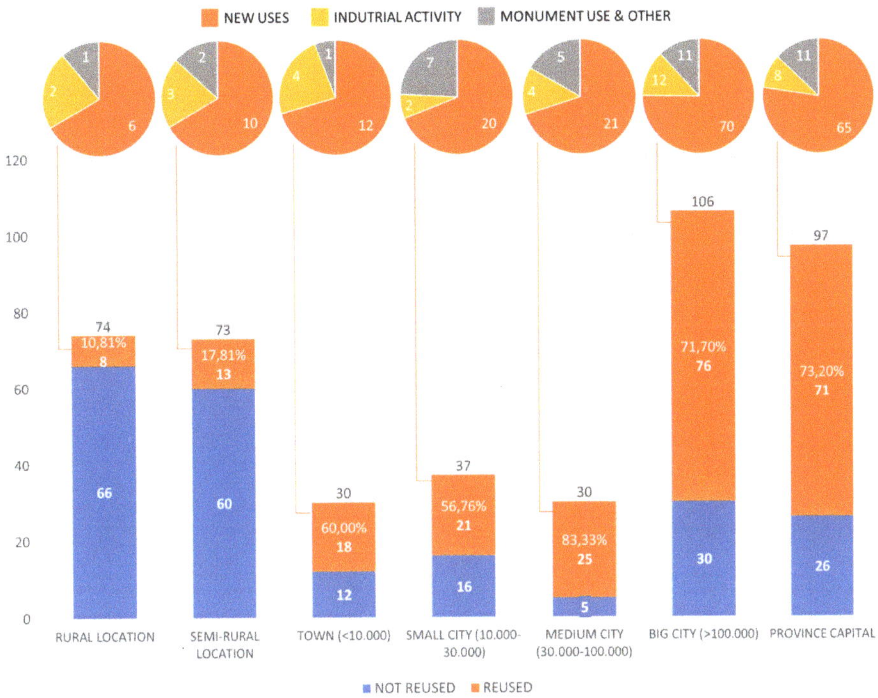

Figure 14. Assets reused and not reused for each classification category referred to the relation with the local population. New uses are of a general nature.

4. Conclusions

The analyses shown in this work are just some of the possible approaches that the developed catalogue allows, but they are considered as representative of the possibilities and potential that these kinds of tools can have in the study, promotion, management, and protection of the assets of the industrial heritage. The results obtained are considered of interest for all of these tasks.

The importance of the incorporation of sufficiently complete criteria structures to the cataloguing initiatives is a key aspect for their usefulness. Industrial heritage cannot be promoted or studied through simple compilations of assets. Their nature and main characteristics must be ordered through an appropriate classification criteria structure that is able to allow both the characterization of the typology and the contextualization of each asset within the sample. This work defines and proposes a full criteria structure for the classification of this type of assets. The main aspects to be taken into account when industrial assets are analyzed are studied through different groups of criteria that are focused on each one of those key aspects. Application guidelines are also given in those cases in which more doubts can rise, what is critical if the catalogue wants to be understood as a real tool.

Thus, the authors consider the developed catalogue as a useful tool in the field of the industrial heritage, both in the international context as a methodological approach and in the particular context of Spain as a tool that represents a significant progress from the previous situation. This usefulness is highlighted in the other works of the authors in which the catalogue and the criteria structure applied to all of them is the substrate in which other methodologies are based [1,19]. Indeed, the new criteria have been included in the structure of the catalogue as a response to the needs that have been identified during its use.

Regarding further developments of the catalogue, and particularly with regard to the actualization of the sample in other territories and the continue revision of the full set of assets considered, the most valuable strategy is a collaborative growth of the catalogue. The authors have previous experiences in the collaborative cataloguing of industrial assets within a university context with positive results. Thus, transforming the catalogue developed in an open tool represents an interesting challenge not only for social promotion of the industrial heritage in Spain, but also as a growth strategy of the sample. Probably, a collaborative identification of assets is the only sustainable and successful strategy for the development of a complete catalogue of this typology in such a large territory as Spain. By availing the knowledge of the potential users of the catalogue located throughout the national territory, which are who know the best examples in their territories, it would be possible to identify and include a large number of assets of great value.

Author Contributions: Conceptualization, J.C. and M.A.S.; methodology, J.C. and M.A.S.; software, J.C. and A.G.-D.; validation, J.C., A.G.-D., L.S. and M.A.S.; formal analysis, J.C., A.G.-D. and L.S.; investigation, J.C. and A.G.-D., L.S. and M.A.S.; resources, J.C., L.S. and M.A.S.; data curation, J.C., A.G.-D., L.S. and M.A.S.; writing—original draft preparation, J.C. and A.G.-D.; writing—review and editing, J.C., A.G.-D., L.S. and M.A.S.; visualization, J.C. and A.G.-D.; supervision, L.S. and M.A.S.; project administration, J.C. and M.A.S.; funding acquisition, J.C., L.S. and M.A.S.

Funding: This research was partly funded by the Manufacturing Engineering Society (http://www.sif-mes.org/) through a mobility grant given to the first author. This research period was developed in the University of Malaga during 2016 and supervised by Professor Sevilla and Professor Sebastián. The results obtained during this research stay represent a significant part of the support data used in this work.

Conflicts of Interest: The authors declare no conflict of interest.

References

1. Claver, J. Methodology for the Analysis and Interpretation of the Assets of the Spanish Industrial Heritage. Application to the Study of the Assets of the Autonomous Community of Madrid. Ph.D. Thesis, Universidad Nacional de Educación a Distancia, Madrid, Spain, 27 April 2016.
2. Institute of the Cultural Heritage of Spain (IPCE), Ministry of Culture and Sport, Spanish Government. National Plan of Industrial Heritage. Available online: http://www.mecd.gob.es/planes-nacionales/dam/jcr:eba404cd-e170-419d-b46a-e241ebd1b1b0/04-texto-2016-pnpi-plan-y-anexos.pdf (accessed on 11 September 2018).
3. Biel, P.; Cueto, G. *100 Elementos del Patrimonio Industrial en España*; TICCIH-IPCE-CICEES: Madrid, Spain, 2009; 324p, ISBN 978-84-937738-6-1.
4. Tostóes, A.; García, C.; Landrove, S. *La Arquitectura de la Industria, 1925–1965*; Registro DOCOMOMO Ibérico; DOCOMOMO Ibérico Foundation: Barcelona, Spain, 2005; 276p, ISBN 84-609-1196-9.
5. The Declaration of Amsterdam. *Adopted at the Congress on the European Architectural Heritag*; The Declaration of Amsterdam: Amsterdam, The Netherlands, 1975.
6. TICCIH. *The Nizhny Tagil Charter for the Industrial Heritage*; Adopted at the Triennial National Assembly of TICCIH; TICCIH: Moscow, Russia, 2003.
7. Llamas, J.; Lerones, P.M.; Medina, R.; Zalama, E.; Gómez-García-Bermejo, J. Classification of Architectural Heritage Images Using Deep Learning Techniques. *Appl. Sci.* **2017**, *7*, 992. [CrossRef]
8. Belhi, A.; Bouras, A.; Foufou, S. Leveraging Known Data for Missing Label Prediction in Cultural Heritage Context. *Appl. Sci.* **2018**, *8*, 1768. [CrossRef]
9. Oglethorpe, M.; McDonald, M. Recording and documentation. In *Industrial Heritage Re-Toleed*; Douet, J., Ed.; Routledge: New York, NY, USA, 2013; pp. 55–62.
10. Rossnes, G. Process recording. In *Industrial Heritage Re-Toleed*; Douet, J., Ed.; Routledge: New York, NY, USA, 2013; pp. 63–69.
11. Claver, J.; Sebastián, M.A. Methodological study and characterization of the industrial heritage of the Autonomous Community of Galicia. *Procedia Manuf.* **2017**, *13*, 1305–1311. [CrossRef]
12. Claver, J.; Sebastián, M.A.; Marcos-Bárcena, M.; Sevilla, L. Identification and characterization of the industrial heritage of the Province of Cádiz. Application of methodologies of identification and classification of assets. In Proceedings of the 24th CUIEET, Cádiz, Spain, 21–23 September 2016.

13. Claver, J.; Sebastián, M.A. Methodological proposes for the study of industrial assets. Application to the Province of Huelva. In Proceedings of the 1st International Congress on Industrial Heritage and Public Work, Huelva, Spain, 26–28 October 2016.
14. Claver, J.; Sebastián, M.A. Methodological analysis of the assets of the industrial heritage in the Autonomous Community of the Region of Murcia. In Proceedings of the 20th International Congress on Project Management and Engineering, Cartagena, Spain, 13–15 July 2016.
15. Claver, J.; Sebastián, M.A.; Sevilla, L. Methodology for the study of the industrial heritage. Application to the Autonomous Community of Andalusia. *DYNA* **2016**, *91*, 136–139. [CrossRef]
16. Rojas-Sola, J.I.; Castro-García, M.; Carranza-Canadas, M.P. Content Management System incorporated in a virtual museum hosting. *J. Cult. Herit.* **2011**, *12*, 74–81. [CrossRef]
17. Rojas-Sola, J.I.; Gómez, M.C.; Castro, M. Windmills in Andalusia: new tools for their valorization. *Asociación de Geógrafos Españoles* **2013**, *62*, 403–427. [CrossRef]
18. Vacca, G.; Fiorino, D.R.; Pili, D. A Spatial Information System (SIS) for the Architectural and Cultural Heritage of Sardinia (Italy). *ISPRS Int. J. Geo-Inf.* **2018**, *7*, 49.
19. Claver, J.; García-Domínguez, A.; Sebastián, M.A. Decision-Making Methodologies for Reuse of Industrial Assets. *Complexity* **2018**, *2018*, 4070496. [CrossRef]
20. Nocca, F. The Role of Cultural Heritage in Sustainable Development: Multidimensional Indicators as Decision-Making Tool. *Sustainability* **2017**, *9*, 1882. [CrossRef]
21. Liu, F.; Zhao, Q.; Yang, Y. An approach to assess the value of industrial heritage based on Dempster-Shafer theory. *J. Cult. Herit.* **2018**, *32*, 210–220. [CrossRef]
22. Giuliani, F.; De Falco, A.; Landi, S.; Bevilacqua, M.G.; Santini, L.; Pecori, S. Reusing grain silos from the 1930s in Italy. A multi-criteria decision analysis for the case of Arezzo. *J. Cult. Herit.* **2018**, *29*, 145–159. [CrossRef]
23. National Institute of Statistical of Spain (INE). Demography and Population Database. 2017. Available online: https://www.ine.es/dyngs/INEbase/es/categoria.htm?c=Estadistica_P&cid=1254734710984 (accessed on 28 October 2018).
24. Google Maps. Available online: https://www.google.com/maps (accessed on 10 January 2018).

Article

Agustín de Betancourt's Double-Acting Steam Engine: Analysis through Computer-Aided Engineering

José Ignacio Rojas-Sola [1,*] and Eduardo De la Morena-De la Fuente [2]

1   Department of Engineering Graphics, Design and Projects, University of Jaén, 23071 Jaén, Spain
2   'Engineering Graphics and Industrial Archaeology' Research Group, University of Jaén, 23071 Jaén, Spain; edumorena@gmail.com
*   Correspondence: jirojas@ujaen.es; Tel.: +34-953-212452

Received: 2 October 2018; Accepted: 16 November 2018; Published: 20 November 2018

Abstract: This article analyses the double-acting steam engine designed by Agustín de Betancourt in 1789 and based on the steam engine of James Watt. Its novelty and scientific interest lies in the fact that from the point of view of industrial archaeology and the study of technical historical heritage there is no worldwide study on this invention, which marked a historic milestone in the design of the steam engines of the Industrial Revolution (1760–1840). This underscores the utility and originality of this research. To this end, a study of computer-aided engineering (CAE) was carried out using the parametric software Autodesk Inventor Professional, consisting of a static analysis using the finite-element method of the 3D CAD model of the invention under real operating conditions. The results have shown that the double-acting steam engine was correctly designed considering that the values of the maximum von Mises stress (188.4 MPa) obtained were taken away from the elastic limit of the material it was made of (cast iron), as well as to the maximum deformations (0.14% with respect to its length) obtained in the same element that presents the maximum stress (opening axle of the high pressure steam valve). Similarly, the maximum displacements (18.74 mm) are located in the mobile counterweights that transmit certain inertia to facilitate the opening and closing of the valves. Finally, if we look at the results of the safety coefficient, whose lowest value was 4.02, we could say that the invention was oversized, following constructive criteria of the time, as there were no resistance tests on materials that would help in the optimization of the design of the invention.

Keywords: Agustín de Betancourt; double-acting steam engine; Autodesk Inventor Professional; computer-aided engineering; mechanical engineering; finite-element analysis; von Mises stresses; displacements; equivalent deformations; safety coefficient

---

1. Introduction

Agustín de Betancourt y Molina was one of the fathers of engineering in the period of the Spanish Enlightenment [1]. This article aims to analyse from the point of view of engineering one of the most controversial inventions of his career, the double-acting steam engine, the first steam engine of its kind to reach the European continent. This invention has already been the object of a detailed study from the graphic engineering point of view, which has allowed us to obtain a reliable 3D CAD model [2] from which the present investigation has been carried out and which led to the discovery of the concept of energy symmetry with which Betancourt designed this invention. Its novelty and scientific interest lies in the fact that from the point of view of industrial archaeology and the study of technical historical heritage there is no worldwide study on this invention, which marked a historic milestone in the design of the steam engines of the Industrial Revolution (1760–1840). This underscores the utility and originality of this research.

Until the end of the eighteenth century, the steam engine known in Europe was that of the English inventor Thomas Newcomen, a simple steam engine that worked thanks to pressure differences in the two chambers of a cylinder. This invention worked from the cooling of the water vapor inside the cylinder, which produced a vacuum. The upper face of the piston, open to the atmosphere, pushed it downward causing it to return to its initial lower position. This movement of the piston operated a rocker arm that impelled other rotating elements through a connecting rod-crank mechanism [3]. Subsequently, James Watt (mechanical engineer and Scottish inventor) worked from 1765 on the Newcomen steam engine, introducing a new element (the condenser) that would triple the performance compared to its predecessor. This simple element allowed advantage to be taken not only of the vacuum produced by the water vapor when condensing but also its expansion, decreasing in this way the amount of water vapor necessary to produce the movement of the piston. He also introduced other elements to increase performance such as the planetary gear system that facilitated the movement of the inertia flywheel, among others.

In 1782 the patent of James Watt reached perfection, when the Scottish engineer adapted the superior part of the cylinder so that the admission of the steam could be realized as much below as above the piston allowing the push of the steam on both its faces.

He also improved the rocker arm designed by Thomas Newcomen. Initially, this rocker arm was attached to the cylinder by means of a chain and therefore only transmitted the movement when the piston moved in the downward direction in the cylinder [3]. However, the double-acting steam engine was attached to the rocker arm by means of a lever that connected the piston shaft to the end of the rocker arm. By means of this connection, the upward stroke of the piston was also exploited but it presented the difficulty of adapting a certain movement of oscillation in the axle of the piston since, while the rocker arm described a circular movement, the axle of the piston moved vertically. Despite the joints (Watt's extended mechanism consisting of an articulated parallelogram) designed to eliminate this movement and ensure an effective rectilinear guidance of the piston, the engine in use presented a significant maladjustment [3].

In September 1785, Betancourt began his second stay in Paris, where he returned with a double commission: to supervise the group of Spanish pensioners and to obtain plans and documents for the Royal Cabinet of Machines of Buen Retiro. During those years, he travelled through different factories and French ports taking notes and making known in Spain a large number of ideas and new techniques with which to stimulate the country's industrial progress. It is during this period that he became aware of the existence of James Watt's steam engine and the enormous progress it represented compared to that of Newcomen [4].

The Spanish engineer, interested in the news about the steam engine, obtained an interview in November 1788 with its inventors James Watt and Matthew Boulton, to be told about their patent. They showed him their factories of buttons and plated silver but none of their steam engines. Even so, he managed to visit the Albion Mills which were being built near the Blackfriars bridge. This installation consisted of three steam engines, one of which was completed [5].

So on December 16, 1789, he presented to the Academy of Sciences of Paris a double-acting steam engine based on the one designed by Watt but improved where it included the theoretical study of the extended mechanism of Watt, as well as solving for the first time in the history of the mechanisms a problem of synthesis of generation of trajectories with three points of precision [6].

As a result of his studies on the steam engine he wrote a memoir on the expansive force of water vapor, which obtained the approval of the Paris Academy of Sciences in September of that same year [7].

There are also two studies on the impact of the Spanish engineer's work on steam engines at the time [8,9] but this invention has never been analysed from the engineering point of view, which highlights the originality and convenience of the present investigation.

The ultimate goal of this study is to perform a static analysis [10] of the double-acting steam engine by the finite-element method [11] under real operating conditions in order to determine whether it was properly sized and would function properly.

2. Materials and Methods

The starting material was only the information available on the website of the Betancourt Project of the Canary Orotava Foundation for the History of Science [12]. Here the information related to the invention is shown, as well as the letter written to his brother José on March 6, 1789 in which he gives news of his steam engine and the report of the Academy of Sciences of Paris on the examination of the invention, signed by Jean-Charles Borda, Mathurin-Jacques Brisson and Gaspard Monge [5].

On the other hand, there are two Betancourt works directly related to this invention. On the one hand in 1790 Betancourt wrote his 'Mémoire sur la force expansive de la vapeur de l'eau' [7], which was one of the first treatises on applied thermodynamics in which the results of the experiments carried out with the double-acting steam machine were shown; and on the other hand, the 'Explication d'une machine destinée à curer les ports de mer' (1808) [13], in which he proposes the design of a mechanical dredger installed on a ship and whose mechanism is propelled by its double- acting steam engine. In this second case, both in the drawings and in the memoir, Betancourt explains the mechanism of the double-acting steam engine, although in less detail.

To obtain the 3D CAD model of the steam engine [2], both the six sheets and the 34-page memory were used, explaining the invention and its operation, which appear in the original Betancourt file [5].

Once the 3D CAD model was obtained the methodology followed for the static analysis object of the present investigation was the same as that used in the study of other Betancourt inventions [14–17], giving it a substantial degree of credibility.

2.1. Operation of the Double-Acting Steam Engine

Although the 3D CAD model of the double-acting steam engine and its operation are perfectly described in the previously referenced publication [2], it has been considered convenient to briefly summarize it in order to facilitate the reader's understanding, due to the high number of components and the complexity of the invention. This explanation is based on two plans along with an indication of the elements that compose the invention (Figures 1 and 2).

Figure 1 represents an isometric view of the set where it can be seen first, a brick building that houses the boiler (18) of the steam engine. This building is not large and fits the models of coal boilers of the time. Secondly, there is a large rocker arm whose balancing axis is supported by two square section columns (the previous one omitted to better see the rest of the elements). Finally, to the left and right of these columns there are two well differentiated parts: on the right, a handle-crank mechanism that moves an inertia flywheel (5) of large dimensions and on the left, the hydro-pneumatic circuit composed of a series of components that regulate its movement.

Through the metal doors of the brick building there is access to a room where the water located in a large boiler over a concave space is heated. In this space coal is burned, causing a thermal plume in the lower zone of the boiler so that the water in the boiler reaches a temperature of 100 °C, transforming it into steam and it leaves the boiler building through an upper pipe A (20).

The steam at high temperature reaches a steam box FF (22) that functions as a double pass valve, that is, one position allows entry to the upper area of the steam cylinder (23) and the other position directs the steam to the lower steam box PQ (16). Thus, the valve system makes it possible to direct the water vapor to the upper or lower face of the piston (27). Valves B and C (44) are correlated to valves D and E (37), so that when the right valve of the upper steam box is open the left valve is closed and in those of the lower steam box the opposite occurs, the right valve closes and the left valve opens.

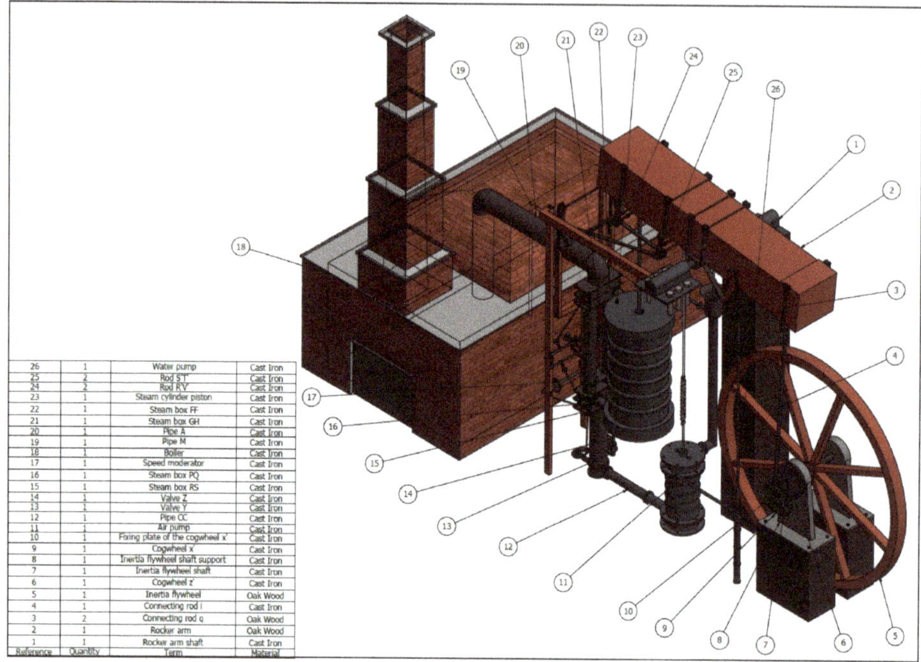

Figure 1. Isometric view of the double-acting steam engine.

Reference	Quantity	Term	Material
26	1	Water pump	Cast Iron
25	2	Rod S'T'	Cast Iron
24	2	Rod R'V'	Cast Iron
23	1	Steam cylinder piston	Cast Iron
22	1	Steam box FF	Cast Iron
21	1	Steam box GH	Cast Iron
20	1	Pipe A	Cast Iron
19	1	Pipe M	Cast Iron
18	1	Boiler	Cast Iron
17	1	Speed moderator	Cast Iron
16	1	Steam box PQ	Cast Iron
15	1	Steam box RS	Cast Iron
14	1	Valve Z	Cast Iron
13	1	Valve Y	Cast Iron
12	1	Pipe CC	Cast Iron
11	1	Air pump	Cast Iron
10	1	Fixing plate of the cogwheel x'	Cast Iron
9	1	Cogwheel x'	Cast Iron
8	1	Inertia flywheel shaft support	Cast Iron
7	1	Inertia flywheel shaft	Cast Iron
6	1	Cogwheel z'	Cast Iron
5	1	Inertia flywheel	Oak Wood
4	1	Connecting rod I	Cast Iron
3	2	Connecting rod q	Oak Wood
2	1	Rocker arm	Oak Wood
1	1	Rocker arm shaft	Cast Iron

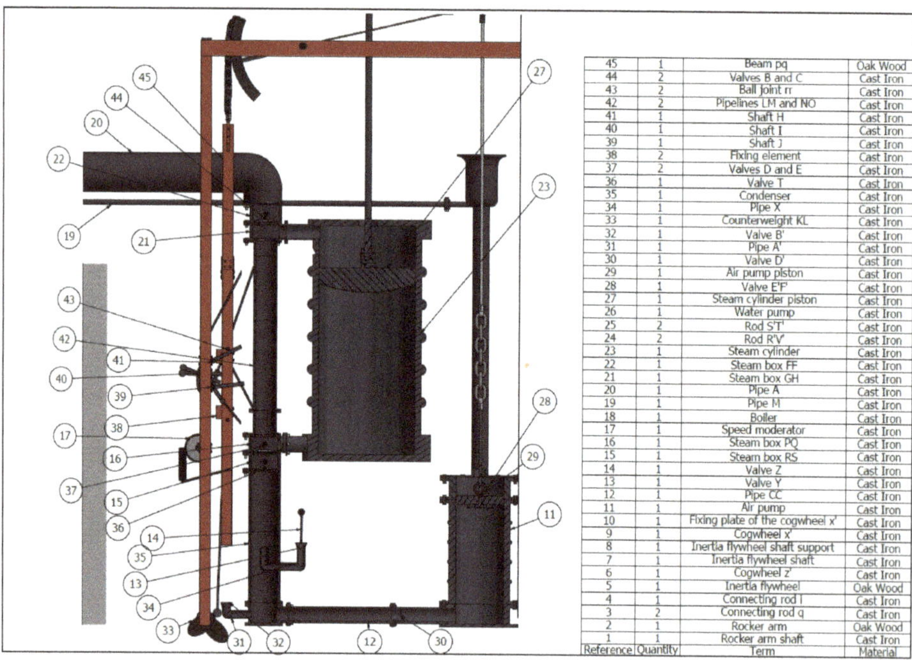

Figure 2. Detail of the plan of the regulating mechanism of the double-acting steam engine.

Reference	Quantity	Term	Material
45	1	Beam pq	Oak Wood
44	2	Valves B and C	Cast Iron
43	2	Ball joint rr	Cast Iron
42	2	Pipelines LM and NO	Cast Iron
41	1	Shaft H	Cast Iron
40	1	Shaft I	Cast Iron
39	1	Shaft J	Cast Iron
38	2	Fixing element	Cast Iron
37	2	Valves D and E	Cast Iron
36	1	Valve T	Cast Iron
35	1	Condenser	Cast Iron
34	1	Pipe X	Cast Iron
33	1	Counterweight KL	Cast Iron
32	1	Valve B'	Cast Iron
31	1	Pipe A'	Cast Iron
30	1	Valve D'	Cast Iron
29	1	Air pump piston	Cast Iron
28	1	Valve E'F'	Cast Iron
27	1	Steam cylinder piston	Cast Iron
26	1	Water pump	Cast Iron
25	2	Rod S'T'	Cast Iron
24	2	Rod R'V'	Cast Iron
23	1	Steam cylinder	Cast Iron
22	1	Steam box FF	Cast Iron
21	1	Steam box GH	Cast Iron
20	1	Pipe A	Cast Iron
19	1	Pipe M	Cast Iron
18	1	Boiler	Cast Iron
17	1	Speed moderator	Cast Iron
16	1	Steam box PQ	Cast Iron
15	1	Steam box RS	Cast Iron
14	1	Valve Z	Cast Iron
13	1	Valve Y	Cast Iron
12	1	Pipe CC	Cast Iron
11	1	Air pump	Cast Iron
10	1	Fixing plate of the cogwheel x'	Cast Iron
9	1	Cogwheel x'	Cast Iron
8	1	Inertia flywheel shaft support	Cast Iron
7	1	Inertia flywheel shaft	Cast Iron
6	1	Cogwheel z'	Cast Iron
5	1	Inertia flywheel	Oak Wood
4	1	Connecting rod I	Cast Iron
3	2	Connecting rod q	Cast Iron
2	1	Rocker arm	Oak Wood
1	1	Rocker arm shaft	Cast Iron

Figure 3 shows the path of the water vapor in the pipes depending on the action of the valves. Figure 3a shows the water vapor at high pressure and temperature impacting on the upper face of the piston of the steam cylinder and causing it to fall, while in Figure 3b the opposite occurs, the water vapor at high pressure and temperature affects the lower face of the piston causing upward movement of the same.

(a)   **(b)**

Figure 3. Movement of water vapor in the hydropneumatic system: (**a**) downward movement of the piston; (**b**) upward movement of the piston.

The incidence of water vapor is not the only cause of the movement of the piston of the steam cylinder. A vacuum is also generated in the condenser (35), creating a remarkable pressure difference which favours its movement. This can be seen in Figure 3b. When the piston is reaching its highest point, water vapor at a lower temperature is dislodged from the cylinder. At that moment valve Y (13) opens, allowing the momentary entry of cold water. This drop in temperature, together with the increase of the space where the steam is located (since the piston of the air pump is rising), produces the condensation of part of the water vapor. Thus, when condensing this steam the pressure decreases locally, becoming lower than the atmospheric pressure and, consequently, both the piston of the steam cylinder and that of the air pump (29) descend.

The piston of the air pump has two small valves, E′ and F′ (28) that allow the upward passage of the water vapor and prevent its return, facilitating also the evacuation of non-condensed water vapor. Also, a last pipe communicates the pump with the atmosphere, presenting at its end a condensation hood so that the water condensed in that hood returns to the boiler for its reuse through a return pipe M (19).

The movement of the piston of the steam cylinder and the downward movement of the piston of the air pump produce the movement of the rocker arm (2). This is connected to an inertia flywheel by a handle-crank mechanism. On one side of the rocker arm the pistons of the steam cylinder and the air pump are connected and at the opposite end there is the connecting rod crank mechanism The connecting rod (4) is connected by means of a joint to a satellite gear (9) and this gear is engaged with a planetary gear (6) solidly connected to the inertia flywheel (5). Finally, the inertia flywheel can move any mechanism.

Finally, it should be noted that there is a water pump, propelled by the same movement of the rocker arm, which serves to flood a space (Figure 3) where are submerged both the air pump and the lower pipe that connects the steam cylinder to the air pump. Thus, this pump maintains the water at a constant level and serves as a water reservoir for use in the condenser.

2.2. Computer-Aided Engineering (CAE)

Based on the 3D CAD model obtained in previous studies [2], reliable results can be obtained in the computer-aided engineering phase, allowing the static analysis to be carried out correctly using Autodesk Inventor Professional software (release 2016, Autodesk, Inc., San Rafael, CA, USA).

2.2.1. Pre-Processing

CAE analysis of the Betancourt double-acting steam engine involves high computational requirements since it consists of a large number of parts subjected to various types of stress. For this reason, it is essential to simplify the model that facilitates the analysis (Figure 4). In a similar manner, the mechanism works in several positions and the valves change noticeably the stresses of the different parts of the mechanism according to whether they are open or closed. Finally, for the static analysis of the engine the two positions in which the valves work have been chosen, since a priori one cannot see which of them is going to have greater stresses.

Figure 4. Simplified model of the double-acting steam engine for static analysis.

In the first place, it has been decided to remove the water boiler and the brick building that houses it since it is not a structural element of the mechanism and its stresses do not affect the rest of the assembly. However, the pipes that enter and leave the building have been taken into account and it is planned to restrict their movements in order to simulate what we would have if this construction existed.

Secondly, all the elements acting as a foundation have been removed, namely the brick supports of the rocker arm and the inertia flywheel. These elements have been replaced by actions that simulate their behaviour. Specifically, the brick supports have been eliminated and on the contrary, the axes supports that allow the movement of the rocker arm and the inertia flywheel are acted upon. Both supports are defined as articulated supports so that the lower faces do not have any degree of freedom but the axis of the supported element is free to rotate.

Thirdly, the last element that is excluded from the simulation is the water tank where the condenser is submerged and the water that cools it, since both elements do not contribute anything structurally to the mechanism and, on the other hand, water cannot be studied in the static analysis. This element is important when explaining the stresses to which the pipes are subjected due to the pressure and temperature differences that it facilitates but its influence does not go any further.

To conclude this section, it can be said that two different positions have been taken for the study based on the opening of the valves of the steam boxes. Valves B and C, as already explained, allow the passage of water vapor at high pressure and temperature to the upper part of the cylinder chamber or to the lower part. In the first position, when valve B is open and valve C is closed, water vapor passes into the PP pipe and enters the lower part of the cylinder pushing the lower face of the piston and, therefore, causing the rocker arm to rise. In the second position, valve C is open and B closed so that water vapor enters directly into the upper part of the cylinder, pushing the upper face of the piston and therefore lowering the rocker arm. The pressures and the vacuum generated in each of them will be considered when preparing their study.

2.2.2. Assignment of Materials

The next stage is to assign material to each of the elements that make up the assembly. However, the original documents of the invention do not specify the materials, although they do show different parts of the machine made of wood, metal or brick.

From the descriptions of other steam engines of the time, it is known what materials were used in those (e.g., the Watt and Boulton steam engine has been widely studied) and according to those specifications and the simple materials that Betancourt would have access to they have been specified. The materials chosen from the library of materials provided by Autodesk Inventor Professional software have been oak, cast iron for metal parts and brick for structural elements. It should be noted that since the structural elements have been excluded from the analysis for the reasons cited in the previous section, the description of the properties of the brick will be omitted.

Autodesk Inventor Professional establishes specific physical characteristics for each material such as thermal, mechanical, elasticity and breaking properties, among others. Cast iron has an isotropic behaviour and its main physical properties are: Young's modulus (120,500 MPa), Poisson's coefficient (0.30), density (7150 kg/m^3) and breaking stress (758 MPa). On the other hand, oak wood has physical properties that depend on the direction in which the elements are studied, since it is an orthotropic material. The most favourable conditions occur when the material works in the direction of the grain since, in the other two orthogonal directions, the physical properties are more limited. For this reason, it is important that in the wooden parts the main axes are always those of the direction of the grain. Thus its main physical properties would be: Young's modulus (9300 MPa), Poisson coefficient (0.0001), density (760 kg/m^3) and breaking stress (46.6 MPa).

2.2.3. Boundary Conditions

Once the materials have been assigned and those elements that cannot be subjected to static analysis eliminated from the simulation, the next stage is to define the boundary conditions of the elements that have a support function. Each support can work in a certain way according to the degrees of freedom that define its mobility, so that the definition of this mobility will affect the static analysis of the complete assembly. The supports can be embedded, articulated, mobile or roller. Thus, the software studies the boundary conditions based on the freedom of movement of each component of the support.

In the first place, the elements that have no freedom of movement are defined, these being the surfaces that are screwed into the brick, such as the supports of the different axles (Figure 5a), as well as those that are geometrically inserted into the brick wall (Figure 5b).

Second, the articulations or elements that rotate freely are defined (Figure 6). These components cannot move longitudinally but can rotate, so they have a lesser degree of freedom. Among them are the hinges of the counterweights, the pulleys and the shaft fastenings.

Finally, it would be necessary to define those surfaces of the support that have freedom in only one direction but in the double-acting steam engine there are none.

On the other hand, the boundary conditions do not depend on the two extreme positions that have been proposed for the static analysis but the contacts between the parts change depending on the position to be studied. Autodesk Inventor Professional automatically detects existing contacts between contiguous surfaces as long as the surface is not excessively complex.

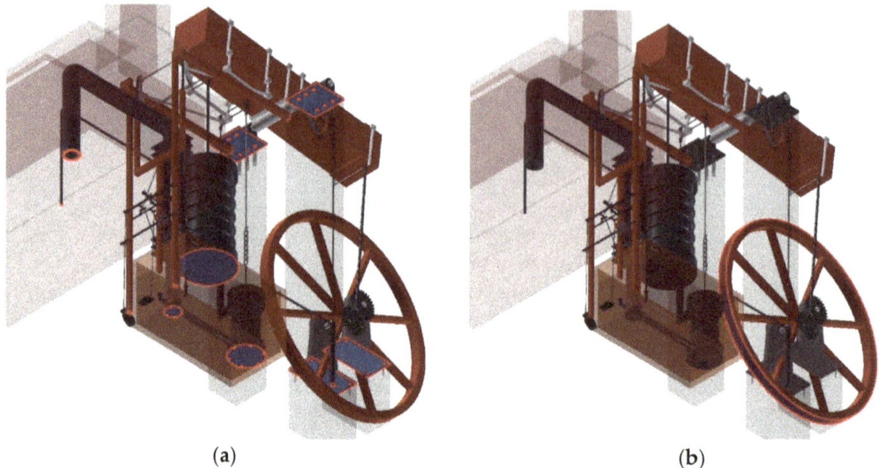

(a)   (b)

Figure 5. Embedded supports: (**a**) fixed elements (**b**) inertia flywheel defined as a built-in element to simulate stop situation.

Figure 6. Rotary supports.

2.2.4. Forces Applied

The next step before performing the simulation is to define the forces acting on the invention, since defining each of them correctly is key to quantifying the stresses that affect the steam engine.

The first of the actions that affects the mechanism is gravity. Autodesk Inventor Professional allows the user to define severity using any direction and magnitude. For modelled engine, it is defined as a generic vector of intensity 9.81 m/s^2 in the direction of the Z axis and in the negative direction. When defined in a generic way, the software represents the gravity vector applied at the centre of gravity of the engine (Figure 7).

Figure 7. Gravitational force applied at the centre of gravity of the engine.

The second action that is going to be characterized is the force that is exerted on the face of the piston of the steam cylinder (Figure 8) and for this the treatise of the English engineer Thomas Tredgold [18] was used, which proposed a method for the calculation of the effect of the strength of the water vapor, as well as its losses and useful work.

(a)   (b)

Figure 8. Force transmitted by the piston: (**a**) downward movement and (**b**) upward movement.

According to Tredgold, the quantifiable losses in the steam engine would be:

(1) The force produced by the movement of steam when entering the steam cylinder: 0.007
(2) Cooling in the steam cylinder and in the pipes: 0.016
(3) Piston friction and losses: 0.125

(4)   The force necessary to expel the steam: 0.007

(5)   The force required to open and close the valves, raise the injection water and overcome the friction of the axles: 0.063

(6)   The loss that comes from intercepting the steam before the end of its trajectory: 0.100

(7)   The force needed to move the air pump: 0.050

Thus, the sum of all losses is 0.368 and therefore the useful energy would be 0.632.

On the other hand, the steam strength in the boiler was generally 900 mm Hg, the temperature of the non-condensed steam 50 °C and its force 100 mm Hg. So it will be necessary to:

$$900 \times 0.632 - 100 = 468.8 \text{ mm Hg} = 0.6373 \text{ kg/cm}^2 \tag{1}$$

Since the pressure on the piston is 0.6373 kg/cm^2 and its area of 2827.43 cm^2, a force will act on it of:

$$F_{piston} = 0.6373 \text{ kg/cm}^2 \times 2827.43 \text{ cm}^2 \times 9.81 \text{ m/s}^2 = 17{,}677 \text{ kN} \tag{2}$$

Moreover, the force on the piston will be located on the lower face when valve B is open and on the upper face when valve C is open.

On the other hand, the pressure received by the face of the piston is the same as that received by the pipes directly connected to the boiler (Figure 9). Thus, depending on which valves are open or closed, the pressure in those pipes can be characterized.

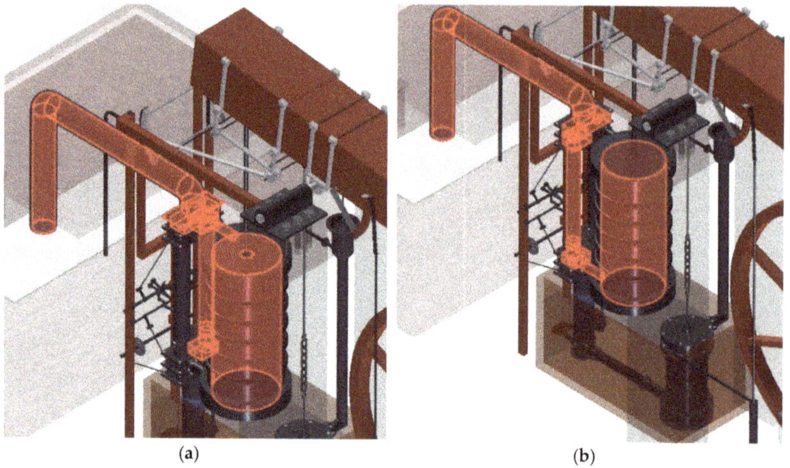

(a)   (b)

Figure 9. Pipes with steam at high pressure and temperature: (**a**) affecting the upper face of the piston and (**b**) affecting the lower face of the piston.

When valve B is open, the pipe coming from the boiler, the upper valve box, the PP pipe, the lower valve box, the entrance to the steam cylinder and the lower part of it have a pressure of 0.6373 kg/cm^2.

In a similar manner, when valve C is open the upper valve box, the upper entrance of the steam cylinder and the upper part thereof are subjected to a pressure of 0.6373 kg/cm^2. The pressure that comes directly from the boiler is called the upper steam pressure. The calculation of the pressure in the lower pipe section with steam at low pressure and temperature (Figure 10) and in the air pump is somewhat more complicated. The air with water vapor that leaves the cylinder does so at a pressure of 0.6373 kg/cm^2 but when it reaches the CC' pipe the water vapor is mixed with a small amount of the water in the tank that is at a lower temperature and higher pressure, causing an immediate condensation of part of the water vapor and therefore producing the vacuum.

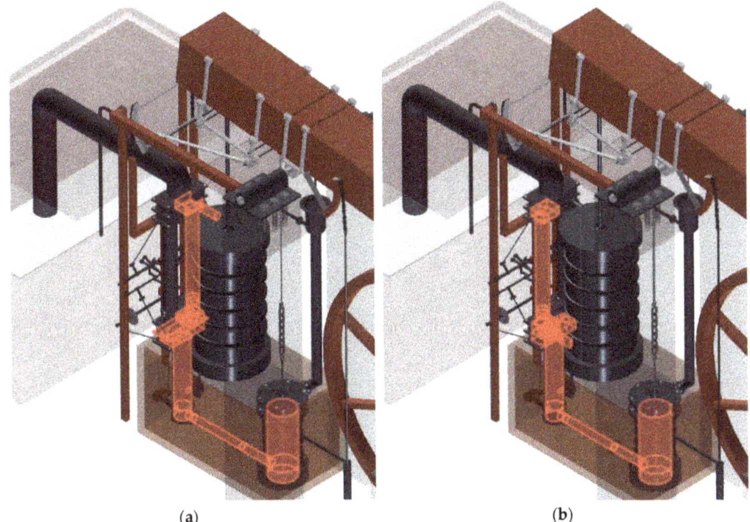

(a)   (b)

Figure 10. Lower pipe with steam at low pressure and temperature: (**a**) subjected to atmospheric pressure and (**b**) subjected to a pressure below atmospheric pressure (by condensation).

Similarly, the pressure outside the pipeline is close to the atmospheric pressure since the pipe is submerged at 0.75 m, so when it is submerged at this depth the external pressure will be:

$$P = 1000 \text{ kg/m}^3 \times 9.81 \text{ m/s}^2 \times 0.75 = 7357.5 \text{ Pa} \tag{3}$$

This pressure is equivalent to 0.0725 atm which, added to the atmospheric pressure, results in an external pressure of 1.0725 atm. On the other hand, the vacuum pressure starts at 0.2960 atm, so the difference in pressure will be that required by the submerged water pump and piping, that is:

$$1.0725 - 0.2960 = 0.7765 \text{ atm, equivalent to } 0.8023 \text{ kg/cm}^2 = 78{,}678.8 \text{ Pa} \tag{4}$$

Finally, the piston of the air pump moves due to the pressure difference between the pipe and atmospheric pressure, so it will be necessary to:

$$1 - 0.2960 = 0.704 \text{ atm, equivalent to } 0.727 \text{ kg/cm}^2 = 71{,}294.3 \text{ Pa} \tag{5}$$

2.2.5. Meshing

Discretization is the last stage before executing the stress analysis of the invention and its object is to obtain a mesh that realistically fits its geometry. As a rule, the greater the density of the mesh, the better it adjusts to its geometry. However, smaller elements need a smaller mesh size than larger ones, although this rule admits some exceptions in the case of complex geometries. Similarly, in places where a specific force is applied it is advisable to establish a higher density mesh, because if the geometry of these points is distorted the results suffer important alterations.

The software used (Autodesk Inventor Professional) presents the option of obtaining the automatic meshing of the part by adjusting some variables in a simple way (Figure 11). By default, this software establishes a mesh formed by tetrahedrons whose average size is 10% of the length of the element, a minimum size of the tetrahedron of 20% of the average size, a maximum variation between tetrahedrons of 1.5 and a maximum angle of rotation of 60°. In the case of the present invention these values are acceptable, although a mesh of higher density will be necessary in the chain links and smaller elements.

Figure 11. Automatic mesh obtained from the double-acting steam engine.

In order to modify the automatic discretization it is necessary to refine the mesh on some surfaces, directly indicating the size of the tetrahedron side. This is the procedure with small elements such as valves, valve axles and screws (Figure 12). On the other hand, the software presents a serious drawback with the chain links that appear in some elements of the steam engine since these links, defined individually by sweeping a circular sector on a closed curve, have a toric geometry, meaning that the assembly of these links is carried out by defining a contact between a point of the surface of the upper link and another of the lower one.

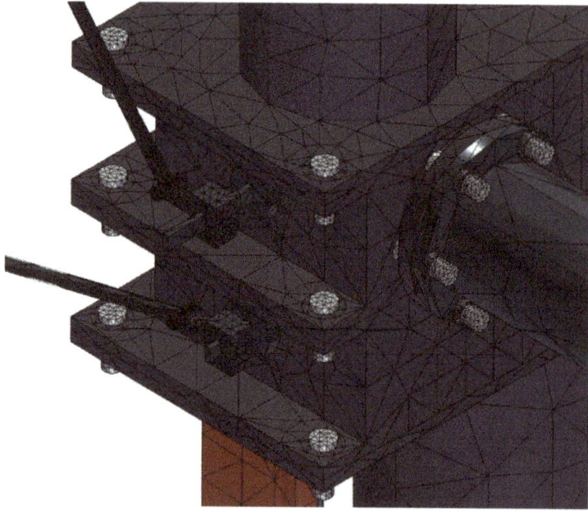

Figure 12. Refinement of the mesh in areas of complex geometry.

In a similar way, the density of the mesh in the zones of contact between links is very large and the contact is established as a single point, so that all the stress is applied to an infinitesimal surface unit, obtaining enormous pressures at these points and therefore distorting the stress results. In order to control and reduce this error at these points the manual mesh control allows the operator to take a mesh density lower than the one established by default, achieving stresses more in line with the real capacity of the chain.

In terms of finite-element analysis with Autodesk Inventor Professional, a convergence criterion has been established since this analysis has been carried out by iterative methods and for a maximum number of ten cycles. Thus, the software compares the result with that of the previous cycle and if that result varies more than 5%, reiterates. However, if the difference is less than 5% the analysis is stopped, adopting the result as definitive. In this study, taking into account computational resources, the analysis used a mesh size of 1,924,288 elements and 3,353,725 nodes.

3. Results and Discussion

Before showing the results of the static analysis of the double-acting steam engine (stress distribution, safety coefficients, deformations and displacements), it is convenient to perform a modal analysis of the engine to determine if there exist any rigid body modes.

Autodesk Inventor Professional performs this simulation by subjecting the structure to vibrations at different frequencies. If the modal frequencies obtained in the analysis are close to 0 Hz then the element to be studied behaves as a mechanism and therefore it would not make sense to perform a static analysis on it.

The eight modal frequencies obtained for the steam engine are: F1: 0.60 Hz, F2: 0.63 Hz, F3: 0.74 Hz, F4: 0.76 Hz, F5: 2.21 Hz, F6: 2.31 Hz, F7: 3.86 Hz and F8: 4.32 Hz. The simulation shows that the first four (slightly lower) frequencies cause displacements in two free counterweights that the invention has whose function is to transmit certain inertia that facilitates the opening and closing of the valves. These counterweights could be excluded in the simulation for a static analysis but this exclusion would affect the solicitation that affects the opening valves of the steam boxes. The dynamism of these elements will therefore be considered so that they do not contaminate interpretation of the results.

The static analysis of the invention has contemplated the study of the two cases indicated above: when the piston of the steam cylinder follows a downward movement and when it moves following an upward movement.

A mesh convergence study has also been performed in order to establish the credibility of the results, since the high stresses are concentrated in very narrow regions of the mechanism. As explained in Section 2.2.5, the discretization of each of the pieces directly affects the results of von Mises stress analysis. The software used allows a refinement of the mesh according to the places where the stress is greater. This process is cyclical since once the regions where the stress is greater are determined, the mesh is refined and the von Mises stress is recalculated. In addition, there are some convergence criteria and in the present study it has been defined that the maximum number of cycles is 10, specifying that when the difference between results is less than 5% the refining process of the mesh is stopped.

Figure 13 shows the convergence curve for the two cases under study. When the piston moves downward, the convergence rate is 4.373% in the fifth iteration (Figure 13a). On the other hand, when the piston moves upwards the convergence rate is lower and therefore more reliable, with a value of 0.013%, although this result is obtained in the seventh iteration (Figure 13b).

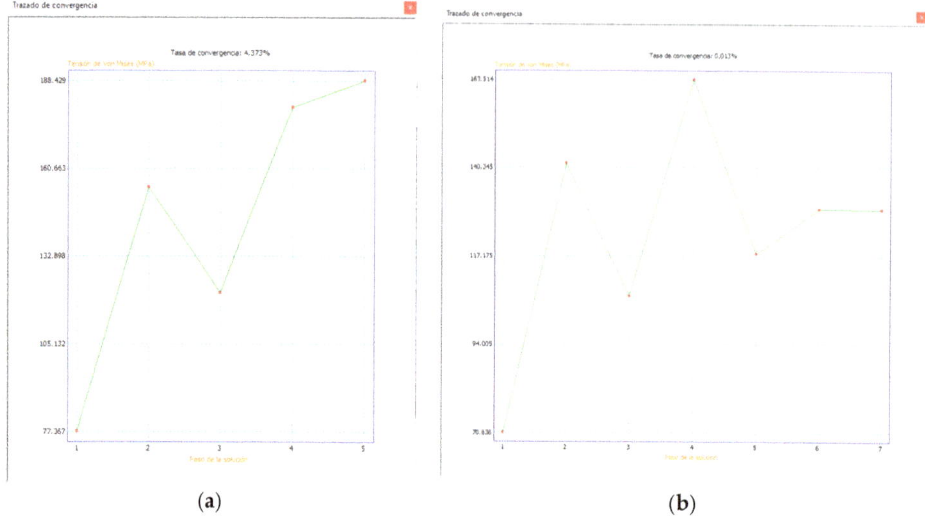

(a)   (b)

Figure 13. Convergence curve: (**a**) downward movement and (**b**) upward movement.

The analysis shows that von Mises stresses are generally low, not reaching 5 MPa (Figure 14), although there are a series of singular points where the stress is higher. This is the case of the opening axle of valve D, one of the valves that diverts water vapor into the piston or into the condensation pipe and occurs when the piston descends, reaching a value of 188.4 MPa (Figure 14a). Also, when the piston rises the maximum stress is located at the same point with a somewhat lower value of 129.6 MPa (Figure 14b). Although these values are high they are not too high, considering the elasticity limit of the material with which the piece is made (cast iron). Figure 15 shows in more detail the point at which the maximum load in the upward direction of the piston is recorded.

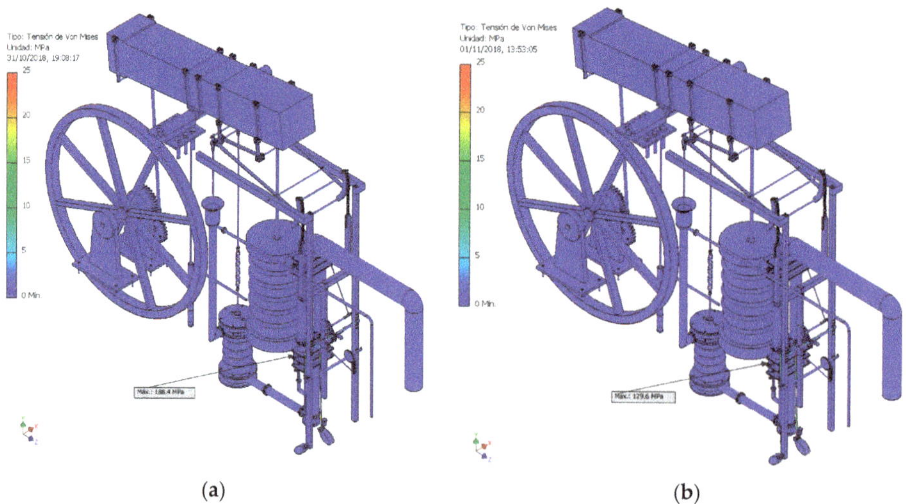

(a)   (b)

Figure 14. Von Mises stress distribution: (**a**) downward movement and (**b**) upward movement.

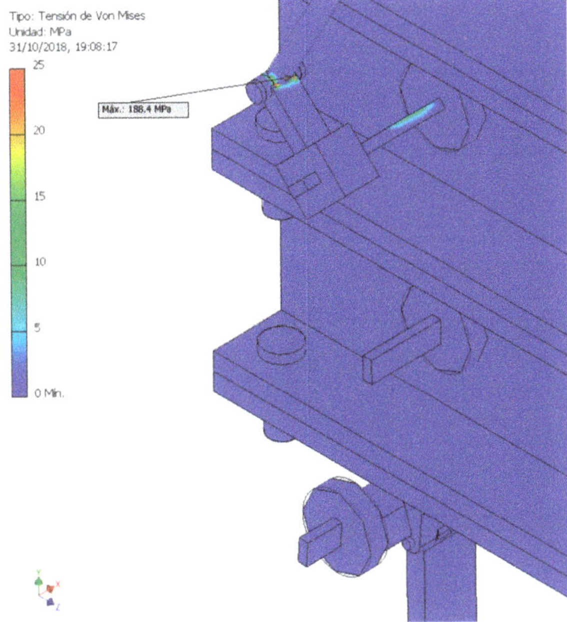

Figure 15. Location of the point where the von Mises stress is maximum.

If the parts that regulate the opening of the valves are excluded, the next set of parts subjected to higher stresses are the rods that join the parallelogram to the piston of the steam cylinder. In Figure 16 this greater stress is located specifically in the second rod, just at the point of insertion of the rod with the frame that serves as support. The von Mises stress value for that point is 47.45 MPa (Figure 16a) when the piston of the steam cylinder moves in a downward direction and somewhat higher with a value of 47.57 MPa (Figure 16b), when moving in an ascending direction, which on the other hand makes sense. The values of this second main stress are already relatively low for the metallic materials with which they are made.

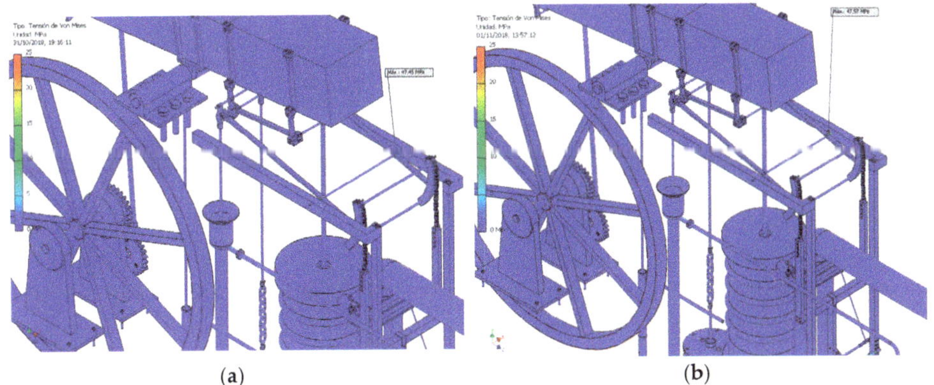

(a)   (b)

Figure 16. Location of the second highest von Mises stress: (**a**) downward movement and (**b**) upward movement.

Another aspect to be studied is the safety coefficient, which is defined as the relationship between the stress to which a part is subjected and the elasticity limit of the material with which it is manufactured. This parameter shows which elements of a structure work with stresses close to the elastic limit of the material and therefore run the risk of breaking and which elements work below it within a particular safety threshold.

Currently, the parts function with a safety coefficient with values between 2 and 4. Parts that work below 2 are too close to the limit of elasticity and suffer significant fatigue, while if the value is above 4, the pieces work far from that limit and therefore are oversized.

In the time of Agustín de Betancourt, the knowledge that existed on the resistance of materials was not very broad and in addition tests were not realized to determine the limits of elasticity of the materials, the reason why mechanisms were generally quite over-dimensioned. The double-acting steam engine is no exception to this rule.

Figure 17 shows the distribution of safety coefficients, it being possible to observe that almost all the elements of the invention have a safety coefficient above 12 and that only a few elements work within a smaller range of values but in any case well over 4.

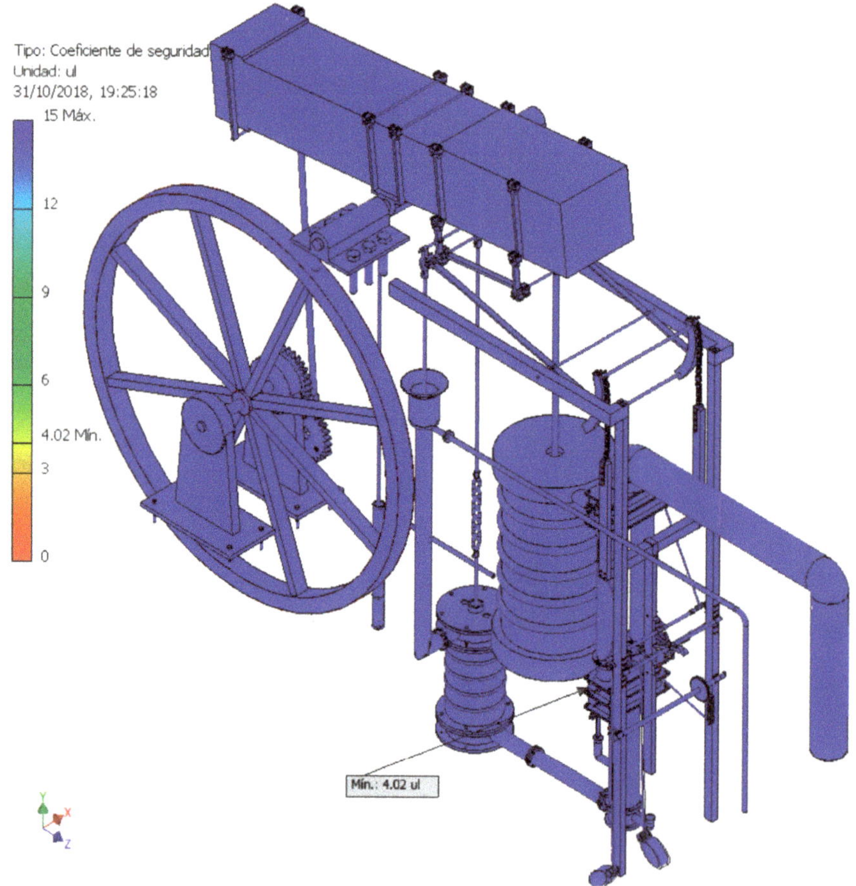

Figure 17. Distribution of safety coefficients.

Furthermore, the point that gives a lower value for the safety coefficient is the opening axle of valve D. The detailed study of the safety coefficient in that axle shows that valve D works with greater stress when the piston of the cylinder is descending and therefore closed and preventing the passage of steam at high pressure to the lower chamber. In this situation the valve axle has a minimum safety coefficient of 4.02 (Figure 18a), above the optimum working values. Similarly, when the valve is open allowing the passage of steam at high pressure the safety coefficient of the valve axle is greater with a value of 5.85 (Figure 18b).

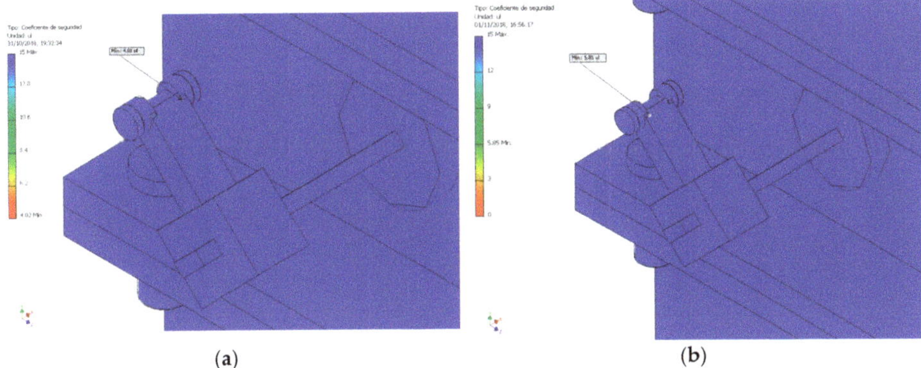

(a)   (b)

Figure 18. Location of the lowest safety coefficient: (**a**) downward movement and (**b**) upward movement.

As indicated previously, if a study is performed excluding the valves the element with the lowest safety factor is the engine speed regulator, with a value of 8.67 when the piston of the steam cylinder falls (Figure 19a) and another of 8.62 when it ascends (Figure 19b). Thus, since the rest of the elements have higher coefficients it is completely clear that the engine is largely oversized.

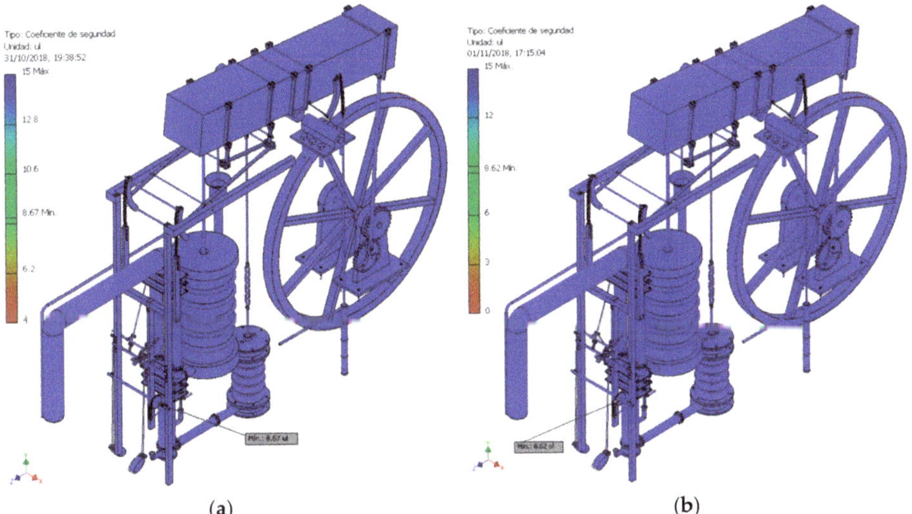

(a)   (b)

Figure 19. Second lowest safety coefficient: (**a**) downward movement and (**b**) upward movement.

On the other hand, the study of the deformation of the elements that make up a mechanism is important, since even if an element does not work in a range of stresses close to the elastic limit of the material, due to its slenderness it can deform its geometry excessively, compromising the correct contact between these elements. Autodesk Inventor Professional shows the equivalent deformation of each element as a relationship between the deformation of the element and its length. In the present study, the maximum deformation is located in the element that suffers the highest stress, that is the opening axle of valve D. However, its deformation is 0.14% with respect to the size of the element when the piston descends (Figure 20a) and 0.10% when it rises (Figure 20b), so it can be considered negligible.

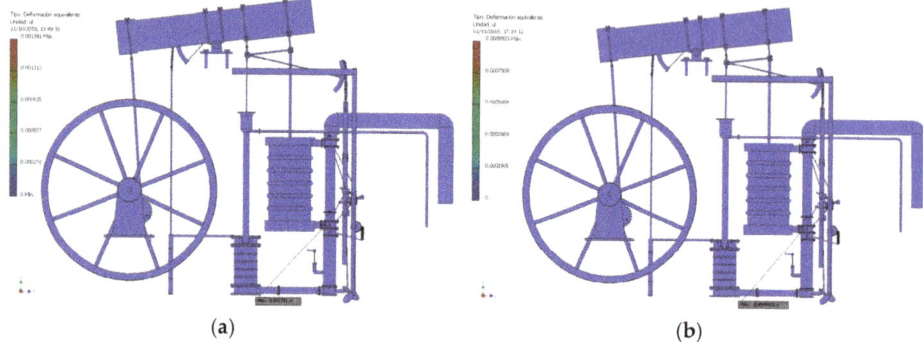

Figure 20. Equivalent deformations: (**a**) downward movement and (**b**) upward movement.

Finally, we should analyse the displacements of some singular elements such as mobile counterweights, which have the highest values. This aspect was indicated previously when carrying out the modal analysis of the invention, since when the counterweights had an inertial function they suffered the greatest displacements.

Thus, when the piston of the steam cylinder moves in a downward direction the counterweight will undergo a displacement of 18.73 mm (Figure 21a) and when it moves in an upward direction of 18.74 mm (Figure 21b). Not in vain, both the material with which the counterweights were made and the elliptical design of the same show that they were designed to bear the wear caused by the continuous movement to which they were subjected.

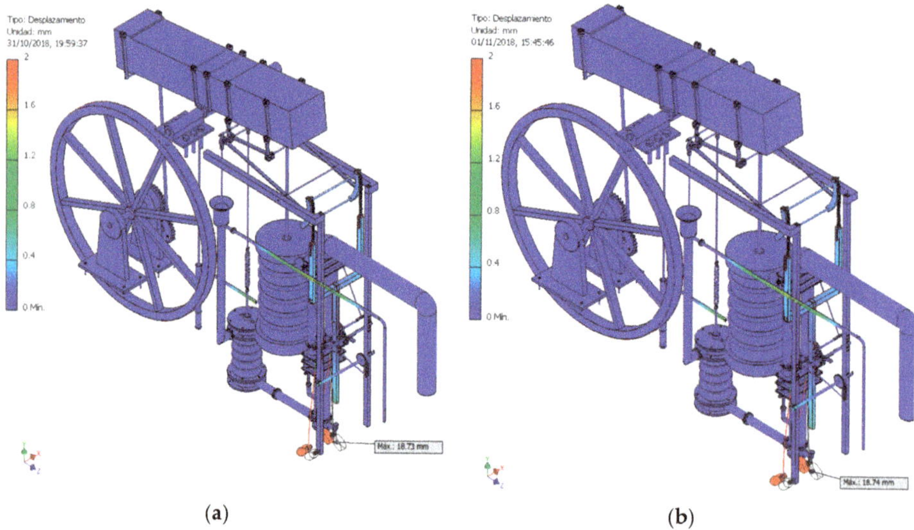

Figure 21. Displacements: (**a**) downward movement and (**b**) upward movement.

4. Conclusions

- The article shows the results of the static analysis carried out on the 3D model of the steam engine designed by Agustín de Betancourt and based on James Watt's double-acting steam engine that he saw working in 1788. For this, CAE techniques have been used thanks to the parametric software Autodesk Inventor Professional.

- The machine presents substantial differences with respect to the model of the Scottish engineer, which makes it a more efficient engine. Its design better transforms the vertical movement of the piston of the steam cylinder into the movement of the rocker arm, using a chain system coupled to the articulated parallelogram invented by Watt in 1784. In addition, the regulating mechanism causes the vacuum generated by the condensation of the water in the pipes to be better used by the engine in the double movement of the piston.

- On the other hand, the results of the static analysis indicate that the double-acting steam engine was completely feasible (it is historically known that Betancourt started one with the Périer brothers in 1790 for a mill in Paris). The study of the von Mises stresses indicates that the point that undergoes a greater tension, the axle of the valve D that regulates the admission of water vapor in the cylinder, presents a value of 188.4 MPa, said value being far from the elastic limit of the material in which it is manufactured (758 MPa). The rest of the pieces work well below this stress value, so it can be said that the choice of materials and dimensions was correct.

- Therefore, according to the results obtained from the safety coefficient it can be said that the structure is a very solid structure, in line with the constructive criteria of the time. The lowest safety coefficient (4.02) is found on the opening axle of valve D, suffering a deformation of 0.14% with respect to its length. Moreover, the maximum displacement (18.74 mm) is found in the mobile counterweights of the engine, mechanisms designed to help in the opening and closing of the valves and designed not to suffer too much wear due to their continuous movement.

- All these data confirm that the invention was correctly sized, although taking into account the values of the safety coefficient of many other elements it could be said that it was clearly oversized. This confirms that practically all the inventions of the time were, due to the fact that there were no resistance tests of materials that would have helped in the design of these inventions.

Author Contributions: Formal analysis (J.I.R.-S. and E.d-.l.M.-D.l.F.); Funding acquisition (J.I.R.-S.); Investigation (J.I.R.-S.); Methodology (J.I.R.-S. and E.D.l.M.-D.l.F.); Project administration (J.I.R.-S.); Supervision (J.I.R.-S.); Validation (E.D.l.M.-D.l.F.); Visualization (J.I.R.-S.); Writing—original draft (J.I.R.-S. and E.D.l.M.-D.l.F.); Writing—review & editing (J.I.R.-S. and E.D.l.M.-D.l.F.).

Funding: This research was funded by the Spanish Ministry of Economic Affairs and Competitiveness (MINECO), under the Spanish Plan of Scientific and Technical Research and Innovation (2013–2016) and European Fund Regional Development (FEDER) under grant number [HAR2015-63503-P].

Acknowledgments: We are very grateful to the Fundación Canaria Orotava de Historia de la Ciencia for permission to use the material of Project Betancourt available at their website. Also, we sincerely appreciate the work of the reviewers of this article.

Conflicts of Interest: The authors declare no conflict of interest.

References

1. Cioranescu, A. *Agustín de Betancourt: Su obra Técnica y Científica*; Instituto de Estudios Canarios: La Laguna de Tenerife, Spain, 1965. (In Spainsh)

2. Rojas-Sola, J.I.; Galán-Moral, B.; De la Morena-De la Fuente, E. Agustín de Betancourt's double-acting steam engine: Geometric modeling and virtual reconstruction. *Symmetry* **2018**, *10*, 351. [CrossRef]

3. Ewbank, T. *A Descriptive and Historical Account of Hydraulic and Other Machines for Raising Water, Ancient and Modern: With Observations on Various Subjects Connected with Mechanic Arts: Including the Progressive Development of the Steam Engine*; D. Appleton and Company: New York, NY, USA, 1842.

4. García Diego, J.A. Huellas de Agustín de Betancourt en los Archivos Breguet. *Anuario de Estudios Atlánticos* **1975**, *21*, 177–221.

5. Betancourt, A. *Mémoire sur une Machine à vapeur à Double Effet*; Academy of Sciences: París, France, 1789; Available online: http://fundacionorotava.es/pynakes/lise/betan_doubl_fr_01_1789 (accessed on 2 October 2018).

6. Cuadrado-Iglesias, J.I.; Ceccarelli, M. La síntesis de generación de trayectorias en Betancourt. In Proceedings of the XXI Spanish Conference on Mechanical Engineering, Elche, Spain, 9–11 November 2016; pp. 420–427. (In Spanish)

7. Betancourt, A. *Mémoire sur la Force Expansive de la Vapeur de l'eau*; Academy of Sciences: París, France, 1790; Available online: http://fundacionorotava.es/pynakes/lise/betan_memoi_fr_01_1790 (accessed on 2 October 2018).

8. Payen, M.J. Betancourt et l'introduction en France de la machine à vapeur à double effet. *Revue D'histoire des Sciences et de leurs Applications* **1967**, *20*, 187–198. [CrossRef]

9. Egorova, O.V. The First Steam Machine in Cuba: Little-Known Pages of Agustin de Betancourt's Work and Life. In Proceedings of the International Symposium on the History of Machines and Mechanisms, Taiwan, China, 11–14 November 2008; pp. 165–174.

10. Hur, D.J.; Know, S. Fatigue analysis of greenhouse structure under wind load and self-weight. *Appl. Sci.* **2017**, *7*, 1274. [CrossRef]

11. Hsia, S.Y.; Chou, Y.T.; Lu, G.F. Analysis of sheet metal tapping screw fabrication using a finite-element method. *Appl. Sci.* **2016**, *6*, 300. [CrossRef]

12. Proyecto Betancourt. Available online: http://fundacionorotava.es/betancourt (accessed on 2 October 2018).

13. Betancourt, A. *Explication d'une Machine Destinée à Curer les Ports de mer*; Academy of Sciences: París, France, 1808; Available online: http://fundacionorotava.es/pynakes/lise/betan_expli_fr_01_1808 (accessed on 2 October 2018).

14. Rojas-Sola, J.I.; De la Morena-de la Fuente, E. Agustín de Betancourt's Mechanical Dredger in the port of Kronstadt: Analysis through Computer-Aided Engineering. *Appl. Sci.* **2018**, *8*, 1338. [CrossRef]

15. Rojas-Sola, J.I.; De la Morena-de la Fuente, E. Agustín de Betancourt's piston lock: Analysis of its Construction through Computer-Aided Engineering. *Inf. Constr.* **2019**, *71*. [CrossRef]

16. Rojas-Sola, J.I.; De la Morena-de la Fuente, E. Agustín de Betancourt's wind machine for draining marshy ground: Analysis of its construction through computer-aided engineering. *Inf. Constr.* **2018**, *70*, e236. [CrossRef]

17. Rojas-Sola, J.I.; De la Morena-de la Fuente, E. Agustín de Betancourt mil for grinding flint: Analysis by computer-aided engineering. *Dyna* **2018**, *93*, 165–169. [CrossRef]

18. Tredgold, T.E. *Tratado de las Máquinas de vapor y de su Aplicación a la Navegación, Minas, Manufacturas etc.*; Imprenta de D. León Amarita: Madrid, Spain, 1831. (In Spanish)

MDPI

St. Alban-Anlage 66

4052 Basel

Switzerland

Tel. +41 61 683 77 34

Fax +41 61 302 89 18

www.mdpi.com

Applied Sciences Editorial Office

E-mail: applsci@mdpi.com

www.mdpi.com/journal/applsci